MyVote

How to Billionaire-Proof Democracy With Electronic Governance

Zed Starkovich

Publisher's Cataloging-in-Publication Data

Starkovich, Zed.
MyVote : How to Billionaire-Proof Democracy with Electronic Governance / Zed Starkovich.

Includes bibliographical references and index.

Summary: MyVote proposes a comprehensive electronic governance platform that unifies access to government services, modernizes democratic participation through secure digital identity and verified polling, and aims to reduce the influence of billionaires and corporate interests on American democracy while increasing accountability, transparency, and citizen empowerment.

ISBN 979-8-9944996-0-3 (pbk.)
ISBN 979-8-9944996-1-0 (ebook)

Library of Congress Control Number: 2026900563

Democracy—United States. 2. Internet in public administration—United States. 3. Political participation—Technological innovations—United States. 4. Access to government information—United States. 5. Elections—Data processing—United States. I. Title.

Suggested BISAC Subject Headings

POLITICAL SCIENCE / Public Policy / General
POLITICAL SCIENCE / Civics & Citizenship

Suggested Classification

LCC (provisional):
JK1764 .S73 2025 (Democracy & citizen participation in the U.S.)

DDC (provisional):
321.8/0973—dc23 (Democracy—United States)

www.ZStarMedia.com

About the Author

Zed Starkovich is a writer, filmmaker, and audio producer whose work sits at the intersection of independent art and digital democracy. Born to Hippie parents in Palo Alto, California, he studied film and physics at NYU before returning to his home state to work in the entertainment industry and raise a family while complaining about politics.

Alongside his work on-set, Zed has spent decades in audio production—recording, mixing, and editing music, producing, directing, and narrating audiobooks, and telling stories that center everyday people rather than corporate myth-making. An early proponent of the "digital future," he uses accessible tools to create and share stories on modest budgets, with a focus on workers, outsiders, and communities whose stories rarely make it to the big screen.

As a father and independent researcher, Starkovich continues to focus on projects that blend craft, technology, and civic imagination, aiming to change how people see themselves, each other, and our collective history as we move toward a brighter future.

Note

While I wish I had the backing of a billionaire-funded think tank with a team of researchers, editors, and an art department, this has been a one-person project dedicated to promoting the potential of a modern digital democracy.

To simplify the intricate topic of electronic governance and create the most straightforward and concise roadmap for the future, I employed various fact-based Large Language Models (or LLMs) for research, proofreading, fact-checking, indexing, source notes, and structuring the text. As a result, I firmly believe that the ability of LLMs to research, analyze, and fact-check the relentless barrage of corporate propaganda we encounter daily will emerge as the most important feature of "AI" in the near future.

Due to financial, artistic, and time constraints, I also edited the graphics myself using generative elements. If any real artists are interested in joining the project and contributing their expertise to replace any of the artwork, I would be delighted to include them in the next edition.

Table of Contents

Introduction

"Government of the people, by the people, for the people,
shall not perish from the earth."

-Abraham Lincoln
Gettysburg Address (1863)

My mother was an underpaid cook and housekeeper who fought every day to raise me right while battling a lifetime of abuse, illness, and generational trauma. I was five when my parents divorced, and twenty-three when she took her own life at age forty-nine, after years of struggling to survive.

Looking back now, I see not only the slow collapse of her life but also the broader unraveling of our country, and how people like her—hardworking and hurting—slipped through the cracks of a political system that claimed to protect them.

Meanwhile, well-paid pundits and podcasters tell us we live in "the most modern, functional, representative, capitalist utopia in world history." But the reality is a system that returns less and less to the people who pay into it —while pouring billions into tax breaks for billionaires, foreign wars, and the growing militarization of ICE and local police forces to keep those struggling people under control.

After witnessing decades of "trickle-down" promises that never reached those most in need, I've spent my life asking a simple question that cuts through the noise: How can we actually make things better?

As a kid raised on TV, that lead me to another important question "How would the people of Star Trek—and the people of our own future— stay connected to their neighbors, help each other live better lives, and elect representatives that actually *represented* their needs?"

Politicians today casually ignore the needs of the people, handing out tax breaks to billionaires and pouring public money into football stadiums for those same elites, while the residents living in their shadows beg for clean water, a living wage, and affordable health care. This isn't governance—it's moral bankruptcy disguised as policy. The brutal absurdity of austerity politics in our modern oligarchy makes the old cry of "taxation without representation" almost sound quaint. What we face now is far worse: a government openly serving corporate power, where each new generation of politicians arrives already owned, and accountability to the public has all but vanished.

This led me to the real question at hand: Do we need to tear down the entire system, or does democracy just need a better user interface?

This project began during the covid pandemic as a work of political science fiction. I'm a filmmaker by trade and much like Gene Roddenberry imagining a better future where the people of Earth had evolved past their differences and resource wars, I looked back at a lifetime of political disappointment and decided to build a video demo of the modern digital democracy that America needs but doesn't yet have.

The result was the video *MyVote: The Billionaire-Proof Digital Democracy*. It may have taken Alexander Hamilton 6 hours to explain his new form of government, but I'm proud to say I summed up Electronic Governance in under 5 minutes! This book is a deeper look at the roadmap presented in that video. My hope in the face of a corrupt and unresponsive political system is that if I can demonstrate a better way to connect with our government, then surely *more people* with *more money, power, and connections than me* will be inspired to actually build it.

Maybe it won't save the world, but I know the comprehensive e-governance platform described in these pages can start making things better by putting every American in the same virtual town square. Things can only get worse if we just keep shouting into the digital void while getting drowned out, distracted, and divided by billionaire-bots and troll-farms.

This isn't a white paper from a well-funded corporate think tank that will get promoted by the profit-driven distraction algorithms and picked up by the local news or celebrated by the national corporate media. I'm just a regular dad trying to survive the class war by using the tools at my disposal to make the world a little better than I found it. So if you watched the video and thought it was full of "cheap, stolen, and AI graphics"… you're right! So far, this has been a one-man project with the video and this book presented as an invitation to build a better future *together* instead of just giving up and letting the billionaires kill us for profit.

With that in mind, I want to give a nod to the power of fact-based LLMs for deep research, text analysis, and academic citations. This project wouldn't have been possible without the time, money, and manpower multiplier of machine learning, and I firmly believe that the ability to analyze and digest massive amounts of data to produce factual, well-sourced information is the most important feature of the "AI" future.

I have major problems with "AI" stealing art and jobs while creating bland derivative nonsense that distracts us from real life (along with polluting our air and wasting energy), but I also believe we need to use every tool available to fight the overwhelming forces against us, and LLMs might be the best weapon we have to fact-check the onslaught of corporate propaganda and disinformation we are presented with on a daily basis.

I'm less worried about an all-powerful AI of the future enslaving humanity because the billionaires of today have already enslaved us under manufactured scarcity and a high cost of living while stealing trillions of dollars from us in wage theft and corporate tax breaks. Used for the collective good, a super-intelligent research assistant in our pockets could, in fact, democratize information and help destroy the petrochemical-patriarchy by proving beyond any doubt that 10 billionaires controlling the world while a billion poor people starve is NOT as unavoidable and impossible to solve as those billionaires claim it is.

As a Californian and long-time Apple user, I reference both throughout this project as an example of what collaboration between business and government could look like when it actually serves the public good. Many see Apple as just another predatory tech giant, but I argue that it remains more committed to individual privacy, creative empowerment, and user choice than its competitors who profit from distraction, manipulation, and surveillance. Even if Apple only cares about selling iPhones, it is better positioned to benefit from the informed, engaged, and independent citizens of a thriving middle class than authoritarian control of the American people. My hope is that Apple's employees will remember their roots in technology and liberal arts and choose to help build the "Star Trek" future of shared progress rather than the Orwellian nightmare that unchecked power makes inevitable.

To quote Public Enemy and The Isley Brothers: We gotta fight the powers that be!

Zed Starkovich
MyVoteGov.org
myvotegov.substack.com
@MyVoteGov.bsky.social

Chapter 1

Welcome to MyVote

Democracy in Your Pocket, Government at Your Fingertips

"Technology has become an integral part of our lives, but it often distracts us from our neighbors and community instead of helping us unite for the common good."

"Introducing MyVote. The revolutionary mobile app that puts your government under your thumbs. Emergency services and government aid that will help your family thrive are now just a tap away, including veteran services, housing, and employment support, putting all your community resources in one place without the distraction of predatory ads and scams. MyVote also makes paying your taxes easier than ever, allowing you to file online in just a few minutes."

"With MyVote, democracy is no longer limited to polling stations and town halls. It's accessible wherever you are. Empowering people and communities like never before."

- MyVote Demo Video

The Problem: Government Is Supposed to Serve You—So Why Can't You Reach It?

We are living in frightening and confusing times. From pandemics, unemployment, corporate price gouging, and unaffordable health care, to endless foreign wars, and armed Federal troops terrorizing our own neighborhoods and killing the innocent citizens they are supposed to protect. The American people feel helpless and afraid for our collective future when technological advances and worker productivity could provide us all with the food, clothing, shelter, clean water, and cheap energy we need to thrive as a society.

Instead of a government that protects our nation's security and stability from "enemies, foreign and domestic," we only see the rich getting richer from our tax dollars as the middle class gets poorer and those at the bottom of the ladder become more desperate than ever. Emails and calls begging our elected representatives for help are ignored while the billionaire-owned news and social media outlets tell us there's no better way to do things—so we all better just suck it up and stop complaining.

Our government *Of The People* has been corrupted into a government *abusing the people for profit*. This crisis of inaccessible, unresponsive, and unaccountable politicians is destroying the lives of the American people instead of protecting and defending us as they were elected to do.

We pay our taxes. We follow the rules. We show up to vote. In theory, our government exists to serve us. In practice, actually accessing government services feels like navigating a deliberately constructed maze designed to make us give up.

Right now, across America, millions of citizens are entitled to services, benefits, and support that could change their lives—and they have no idea those services exist. Or they know the services exist but can't figure out how to access them. Or they started the application process and got lost in bureaucratic quicksand. Or they're being actively scammed by predators impersonating government services.

Here's what trying to access government looks like for most Americans...

The Emergency Services Nightmare

Your elderly mother falls and breaks her hip. She needs immediate help. You try to find emergency services for seniors in your community. You Google it. You get:

- Predatory ads from private companies charging $200/month for "emergency monitoring" (preying on your panic).
- Seventeen different websites, half of which haven't been updated since 2018.
- Contradictory information about what services are available.
- Phone numbers that go to voicemail during "business hours."
- Scam websites mimicking government services that steal your mother's personal information.
- No clear answer about what actual government services exist, who qualifies, or how to access them.

By the time you navigate this chaos, hours have passed. Your mother is suffering and you are desperate enough to pay the predatory private company, because you can't find the legitimate government program that would provide the service for free or at a fraction of the cost.

This scenario plays out thousands of times daily across America.

The Veterans Services Disgrace

You are a veteran. You served your country. You are entitled to healthcare, disability benefits, education support, housing assistance, and employment services. In theory.

In practice:

- The VA website is a labyrinth of broken links and outdated information.
- Different benefits are managed by different systems that don't talk to each other.
- You have to create seven different accounts with seven different passwords.
- Each application requires submitting the same documents multiple times.

- Wait times for responses stretch into months or years.
- You have no idea what you're entitled to because information is scattered across dozens of uncoordinated sources.
- Private companies run predatory ads claiming they can "help veterans get benefits" (for a fee) when those same benefits should be accessible for free.
- Your medical records are in one system, your disability claim in another, your education benefits in a third—none of them integrated.

Twenty veterans die by suicide every day in America.[1] Many were struggling to access the services they earned. Many didn't know those services existed. Many gave up trying to navigate the bureaucracy.

That is not just a failure. That is a betrayal.

The Housing Support Lottery

You are struggling financially. You can't pay rent. You are facing eviction. You have kids. You are terrified.

You know housing assistance programs exist—Section 8, emergency rental assistance, housing vouchers, transitional housing. You've heard about them. Can you access them?

- Housing assistance programs are run by different agencies at federal, state, and local levels.
- Each has different eligibility requirements, different application processes, different wait lists.
- Information about what programs exist in your area is scattered across multiple websites.
- Many programs are so underfunded that wait lists close entirely— you can't even apply.
- Applications require extensive documentation (tax returns, pay stubs, birth certificates, proof of citizenship) that you may not have readily available.
- You discover programs exist that you could have qualified for months ago—but nobody told you.
- By the time you navigate the system, you've been evicted.

Meanwhile, scammers run ads on Facebook: "Emergency Housing

Assistance - Apply Now!" They steal your personal information and your application fee. The "emergency assistance" never materializes.

More than 770,000 Americans experienced homelessness on any given night in 2024—an 18 percent increase from the previous year.[2] Nearly 150,000 of these were children.[3] Many could have been helped by programs that exist —if they could have found and accessed them in time.

The Employment Support Maze

You lost your job. You need help finding work, retraining for a new career, or starting a business. Government programs exist for all of this—workforce development, job training, small business loans, unemployment insurance, apprenticeship programs.

Can you find them?

- The Department of Labor has programs. So does your state workforce agency. So does your county. So do various federal grant programs. None are coordinated.
- Job training programs have different eligibility requirements—some for displaced workers, some for veterans, some for youth, some for adults, some for specific industries.
- Small business support is split between SBA (federal), state economic development agencies, local chambers of commerce, and various grant programs.
- Unemployment insurance is state-managed but federally supported —good luck figuring out which agency to contact when your claim has issues.
- Information is scattered across dozens of websites with different formats, different applications, different requirements.
- Many programs go unused because people don't know they exist.

You spend weeks trying to figure out what help is available. By then, you are behind on mortgage payments, your savings is depleted, and your stress is overwhelming. If you had known about the job retraining program on day one, you could already be enrolled and on the path to new employment.

Why is finding help harder than losing your job?

The Tax Filing Torture

You have to file taxes. It's not optional. The IRS already knows what you owe (they have all your income information from employers and banks). But you still have to spend hours or hundreds of dollars to tell them what they already know.

The current tax filing system is deliberately complicated:

- Free filing is technically available—but hidden behind a maze of restrictions and limited eligibility.
- Most Americans end up paying TurboTax or H&R Block $50-$200 to file taxes that should be free.
- These companies lobby extensively to keep tax filing complicated so you'll keep paying them.[4]
- The IRS can't build a simple free filing system because Congress, influenced by tax prep company lobbying, won't fund it.[5]
- If you make a mistake, you might owe penalties—even though the government could have just told you what you owed in the first place.
- Small business owners spend dozens of hours on taxes, or pay accountants thousands of dollars, for a process that could be streamlined with modern technology.

Americans collectively spend 6.5 billion hours and over $260 billion annually on tax compliance.[6] Not paying taxes—just figuring out how to report what the government already knows.

It's not an accident that it's this complicated. It is profitable for tax prep companies. It is politically useful for those who want citizens to hate government, and it is completely unnecessary with modern technology.

The Information Overload Paralysis

Even if you overcome all these barriers and finally access a government website, you're confronted with:

Incomprehensible Language: Government websites are written in bureaucratic jargon that requires a law degree to understand. "Pursuant to subsection 42(b)(3) of Title 26, eligible applicants may submit documentation evidencing qualification for supplemental assistance…"

Translation: "If you qualify, send us proof."

Buried Information: The one piece of information you need is hidden seventeen clicks deep, behind three different menus, in a PDF that hasn't been updated since 2019.

Broken Links and Dead Ends: You click "Apply Here" and get a 404 error. You call the help line and get disconnected. You email the contact address and it bounces back.

Contradictory Instructions: The federal website says one thing. The state website says something else. The local office tells you something completely different. All three claim to be authoritative.

Predatory Ads Everywhere: You search for "veteran benefits" and the first five results are paid ads from companies that charge you for services that should be free. Some are legitimate (expensive) services. Some are outright scams. You can't tell the difference.

The human cost compounds daily:

- A diabetic veteran can't find the VA pharmacy system, so he rations his insulin and ends up in the emergency room.
- A single mother doesn't know about WIC (nutrition assistance), so her children go hungry while she's entitled to help.
- A laid-off factory worker doesn't learn about retraining programs until it's too late to enroll.
- An elderly couple gets scammed out of their savings by a fake Medicare hotline because they couldn't find the real number.

These aren't edge cases. These are daily realities for millions of Americans.

And it's completely, entirely, inexcusably unnecessary.

The technology exists to fix all of this. The information exists. The services exist. The funding exists (or should exist, and where it doesn't, we can see that clearly and demand it).

What's missing is a functional system to connect citizens to their government.

The Solution: MyVote—Your Government, Unified and Accessible

MyVote is the single, comprehensive, verified platform that puts every government service, every piece of information, every resource you need in one place, accessible from the device already in your pocket.

Not another government website. Not another bureaucratic portal. A unified civic operating system that fundamentally transforms how you interact with government—making it as easy to access public services as it is to order dinner, check your bank balance, or message a friend.

Here's what MyVote does:

Emergency Services at Your Fingertips

When crisis hits, MyVote is your lifeline.

One Tap Emergency Access:

- Medical emergency services
- Crisis intervention hotlines
- Domestic violence resources
- Mental health crisis support
- Suicide prevention services
- Disaster assistance
- Emergency financial assistance
- Emergency housing/shelter information
- Food banks and meal programs
- Legal aid for emergencies

Geolocation-Aware: MyVote would know where you are and show you the closest, most relevant resources:

- "Nearest emergency shelter: St. Mary's Community Center, 0.8 miles, beds available."
- "24/7 crisis counseling: Call or text, response time under 5 minutes."
- "Food assistance available today: Community Kitchen, 3-6 p.m., no documentation required."

Direct Connection: Not just information—actual access. One tap to call. One tap to text. One tap to get directions. One tap to start an application. No searching. No navigating. No barriers between you and help.

Family Safety Features:

- Share your location with emergency contacts
- Alert family members to resources they might need
- Store critical medical information for emergency responders
- Document safety plans for domestic situations
- Access child protective services information when needed

Disaster Response Integration: When hurricanes, wildfires, floods, earthquakes, or other disasters strike:

- Real-time evacuation information
- Emergency shelter locations and availability
- FEMA assistance application
- Disaster unemployment assistance
- Lost document replacement services
- Missing persons registry
- Volunteer coordination

All verified. All legitimate. Zero scams. Zero predatory ads.

No Ads. No Scams. No Exploitation.

This is critical: MyVote has zero advertising and zero third-party monetization.

When you search for veteran benefits, you see actual veteran benefits—not paid ads from companies that want to charge you 30% of your disability payment to file your claim.

When you look for housing assistance, you see legitimate government programs—not predatory payday lenders offering "fast cash" at 400% APR.

When you need emergency help, you get real resources—not scammers impersonating government services to steal your personal information.

MyVote is funded by the government—by your tax dollars—to serve you without extracting additional money from you. That's how government should work. That is what MyVote delivers.

Veteran Services—Finally Unified

If you served, MyVote ensures you get everything you earned.

Your Complete VA Profile: One dashboard showing:

- Healthcare enrollment and benefits
- Disability rating and claims status
- Education benefits (GI Bill) usage and remaining eligibility
- Home loan eligibility and pre-qualification
- Pension and compensation payment history
- Vocational rehabilitation status
- Dependent and survivor benefits

Integrated Medical Records: Your complete VA medical history, accessible to you and (with your permission) to your healthcare providers. No more repeating your medical history at every appointment. No more lost records.

Claims Tracking:

- File new claims directly through MyVote
- Upload supporting documentation from your phone
- Track claim status in real-time
- Receive notifications of decisions
- Appeal directly through the app if denied
- Connect with VSO (Veteran Service Organization) representatives

Education and Employment:

- GI Bill application and management
- Vocational rehabilitation enrollment
- Job matching based on your military experience
- Resume translation (military skills to civilian job requirements)

- Apprenticeship and training programs
- Small business support for veteran entrepreneurs

Housing and Financial Support:

- VA home loan application and pre-qualification
- Housing adaptation grants (for disability modifications)
- Emergency financial assistance
- Pension programs
- Survivor and dependent benefits

Community Connection:

- Local veteran organizations
- VA facility locations and wait times
- Peer support groups
- Mental health resources specifically for veterans
- Benefits counseling appointments

No more navigating seventeen different systems. No more waiting months to hear about claims. No more predatory companies charging veterans for services that should be free.

Housing and Employment Support For Every Citizen—All in One Place

MyVote consolidates every resource for housing stability and employment success.

Housing Assistance:

- Section 8 / Housing Choice Voucher application and waitlist status
- Emergency rental assistance
- Utility assistance programs
- First-time homebuyer programs
- Down payment assistance
- Foreclosure prevention resources
- Homeless prevention services
- Transitional housing programs

- Public housing applications

Employment Services:

- Job search tools with verified listings (no scams)
- Unemployment insurance application and claims
- Job training and apprenticeship programs
- Career counseling and assessment
- Resume building tools
- Interview preparation resources
- Disability employment support
- Ex-offender re-entry programs
- Immigrant employment services

Small Business Support:

- SBA loan applications
- Business plan templates and counseling
- Licensing and permit information
- Tax ID application
- Mentor matching programs
- Contracting opportunities (government contracts for small businesses)
- Minority and women-owned business certifications

Financial Stability:

- SNAP (food assistance) application
- WIC (nutrition for women, infants, children)
- TANF (temporary assistance for families)
- Child care assistance programs
- Energy assistance (LIHEAP)
- Free tax preparation (VITA program locations)
- Financial counseling resources
- Debt management programs

All services show:

- Clear eligibility requirements ("You qualify" or "You don't qualify because…")
- Required documentation (and help getting it if you don't have it)

- Application status tracking
- Estimated wait times
- Appeal processes if denied
- Local office contact information

Taxes Made Simple—Finally

MyVote transforms tax filing from torture to trivial.

Automatic Pre-Filing: MyVote integrates with IRS systems (through X-Road) and pre-fills your return with information the government already has:

- W-2s from your employer
- 1099s from banks and investments
- Previous year's return information
- Deduction and credit eligibility based on your profile

Guided Filing:

- Plain-language questions (not tax code jargon)
- Automatic calculation of deductions and credits you qualify for
- Warning flags if something looks wrong
- Comparison to previous years to catch errors

Direct Filing—Completely Free:

- No paying TurboTax $200 for something that should cost nothing
- No artificially restricted "free" versions that upsell you constantly
- No lobbying-created complications
- Just file, for free, in minutes

For Simple Returns: If you have standard W-2 income, standard deductions, and no unusual circumstances, filing takes literally 3-5 minutes:

- Review pre-filled information
- Confirm it's correct
- Submit
- Done

For Complex Returns: If you have business income, investments, rental properties, etc., MyVote provides:

- Step-by-step guidance
- Connection to free tax preparation assistance (VITA program)
- Option to consult with tax professionals (but you're not forced to pay for basic filing)

Payment Options:

- Pay balance directly through MyVote (no separate payment portal)
- Set up payment plans if you can't pay in full
- Track refund status in real-time
- Direct deposit of refunds (no waiting for checks)

Historical Records:

- Access all previous tax returns
- Download forms for mortgage applications, financial aid, etc.
- Audit support if needed (rare, but available)

The goal: Tax filing should take minutes, not hours. It should cost nothing, not hundreds of dollars. It should be as easy as checking your bank balance. MyVote makes it happen.

Community Resources—Organized and Accessible

MyVote isn't just federal services—it's every level of government and community support.

Local Services:

- Parks and recreation programs
- Library system (digital card, book checkouts, event calendars)
- Public transportation schedules and passes
- Trash and recycling schedules
- Utility payment and management
- Local permits (parking, events, construction)
- Animal control and pet licensing
- Local health departments

Education:

- Public school enrollment
- School lunch program application
- Special education services
- College financial aid (FAFSA)
- Student loan management
- Scholarship databases
- Adult education and GED programs

Healthcare:

- Medicaid and CHIP enrollment
- Healthcare marketplace (ACA) enrollment
- Medicare enrollment and management
- Prescription assistance programs
- Mental health resources
- Substance abuse treatment locators
- Free and low-cost clinic locations
- Health department services (immunizations, STD testing, etc.)

Legal Services:

- Legal aid eligibility and referrals
- Court information and schedules
- Small claims court filing
- Traffic ticket payment
- Name change and legal document services
- Immigration legal services
- Tenant rights information
- Consumer protection resources

All organized by your location. All customized to your needs. All without predatory ads, scams, or companies trying to charge you for services that should be free.

Democracy Beyond the Ballot Box

MyVote isn't just services—it is civic engagement.

Stay Informed:

- Track legislation that affects you
- Follow your representatives
- Read verified news relevant to your community
- Access public records and historical documents

Participate Actively:

- Sign petitions that matter to you
- Comment on proposed regulations
- Participate in town halls (virtual or in-person)
- Vote in local elections and referendums
- Build your own budget priorities to share with representatives

Hold Representatives Accountable:

- See their voting records
- Track their campaign promises
- Message them directly with concerns
- Rate their responsiveness
- Verify their public statements

Build Community:

- Connect with neighbors on local issues
- Organize around shared concerns
- Share verified information (not misinformation)
- Participate in authenticated discussions (no bots, no trolls)
- Build coalitions for change

Democracy isn't just Election Day. It's every day. MyVote makes that real.

Government That Actually Serves You

For too long, we've accepted that government has to be difficult to navigate, slow to respond, and frustrating to deal with. We've normalized the idea that accessing services you're entitled to requires patience, persistence, and often professional help.

That's not inevitable. That's a choice. A choice made by those who benefit from an inaccessible system—whether private companies profiting from filling the gaps, or political interests who benefit when citizens think "government doesn't work."

My Vote is the choice to do better.

Not by creating more bureaucracy—by cutting through it with technology that already exists.

Not by building a surveillance state—by creating transparent systems you control.

Not by replacing human services—by making those services accessible when and where you need them.

Just by making government work the way it should have worked all along.

Real-World Proof: This Isn't Science Fiction

MyVote is built on X-Road, the secure data exchange layer originally developed in Estonia 25 years ago. It has expanded globally as a leading solution for electronic governance and digital public infrastructure, and is currently deployed in 25 countries. Over 542 million people depend on X-Road infrastructure daily as the backbone for digital services provided by public and private sector organizations.[7]

Here are a few of the biggest examples...

Estonia's X-Road System:

- 99% of government services available digitally[8]
- Tax filing takes 3-5 minutes for 95% of citizens[9]
- Healthcare records unified and accessible

- Birth registration happens automatically (hospitals report to government systems)
- Starting a business takes 15 minutes online
- Digital signatures save 2% of GDP annually[10]
- **Result:** Highest citizen satisfaction with government in Europe

Singapore's SingPass:

- Single digital identity for all government services
- 4.5 million users (in a country of 5.9 million)[11]
- Access to 2,700+ government and private sector services[12]
- Mobile app enables instant authentication and document sharing
- Over 350 million transactions annually[13]
- **Result:** 97% digital government service usage among citizens aged 15+

South Korea's K-Government:

- Unified government service portal serving 50+ million citizens
- Tax filing, healthcare management, business registration, all integrated
- Mobile-first design with 80%+ smartphone penetration
- **Result:** Ranked #1 globally in e-government development

Denmark's NemID:

- Single digital ID used for banking, government, healthcare
- 5.8 million users (in a country of 5.9 million)
- Reduced government administrative costs by 30%
- **Result:** 90%+ citizen satisfaction with digital government

These aren't small countries with easy problems. They are diverse democracies that decided to invest in their digital infrastructure. They have proven it works. MyVote can bring the same success to America.

Why This Matters to You

Time: Hours saved every year *not* navigating bureaucracy, *not* hunting for information, *not* re-submitting the same documents to different agencies.

Money: Hundreds or thousands of dollars saved on tax preparation, avoided scams, and accessed benefits you didn't know existed.

Stress: The crushing anxiety of dealing with government disappears when services are actually accessible and understandable.

Dignity: You're not a supplicant begging bureaucracy for help you're entitled to. You're a citizen accessing services your taxes fund, easily and respectfully.

Opportunity: When you can actually access job training, education benefits, housing support, and small business resources, you can build the life you want instead of being trapped by circumstances.

Safety: Emergency services that are actually accessible in emergencies. Resources that help before crisis becomes catastrophe. Support that's there when you need it.

What Comes Next

The following sections will detail how MyVote achieves this vision.

- **Integration with Existing Systems:** How X-Road connects government databases without centralizing power or compromising security.

- **Authenticated Users and Digital ID:** How biometric verification creates accountability while protecting privacy.

- **Representation and Accountability:** How MyVote gives you direct access to representatives and makes them accountable for responsiveness.

- **Personalized Dashboard:** How your MyVote interface delivers exactly the information and services you need, without noise or manipulation.

- **Verified Polling and Participatory Budgeting:** How MyVote enables continuous democratic participation beyond just voting every two years with instant verified polling to create a data-driven democracy based on factual information our representatives and the profit-driven media can't ignore.

- **Elections and Voting:** How MyVote transforms elections from confusing chaos into informed, transparent, verifiable democratic choice.

- **Community Empowerment:** How MyVote helps citizens and their communities unite for their common good, enabling real, accountable representation from their elected representatives.

Each section builds on this foundation: government services should be as accessible as any other digital service you use daily, but with stronger security, complete transparency, and zero commercial exploitation.

The Bottom Line

One app. All your government services. Emergency resources, veteran benefits, housing support, employment services, and tax filing—all accessible with one tap. No predatory ads. No scams. No bureaucratic nightmares. Just the government you pay for, finally accessible when and where you need it. Democracy isn't just polling stations anymore—it will be in your pocket, empowering you and your community every single day. And unlike the current system—which is broken by design and resistant to change—MyVote will be built to evolve, improve, and serve every citizen.

Many states have already taken their first steps toward the comprehensive system described here, including the great state of Wisconsin, which has been hosting it's election and voting information online at MyVote.wi.gov since 2012. With secure modern technology every citizen of every city and state could have their own MyVote.gov town square.

Summary of Best Practices
How MyVote Changes Everything

One Account, Verified Once: _You create your MyVote account one time, with the biometric verification you already use on your phone everyday. From then on:_

- _Every service recognizes you_
- _Your information pre-fills applications_
- _You control what data is shared with whom_

Mobile-First Design: _MyVote is designed for the device in your pocket:_

- _Clean, simple interface_
- _Fast loading even on slow connections_
- _Works offline for critical features_
- _Accessible design for all abilities_
- _Multiple language support_

Privacy by Default:

- _You control what information is shared_
- _Government agencies can't see your data without your permission_
- _Your activity is private (what services you access, what information you read)_
- _No tracking for advertising_
- _No selling your data_
- _Strong encryption protecting everything_

Integration, Not Replacement: _MyVote doesn't replace existing government systems—it integrates them:_

- _Backend systems stay in place (agencies keep their databases)_
- _X-Road connects them securely_
- _MyVote provides the unified front-end interface_
- _Citizens get simplicity; agencies keep control of their data_

Continuous Improvement:

- *Open-source codebase that anyone can audit*
- *Regular security updates*
- *User feedback integration*
- *New services added continuously as more agencies integrate*
- *Accessibility improvements based on user needs*

Chapter 2

Integration With Existing Systems
Breaking Down the Walls Between You and Your Government

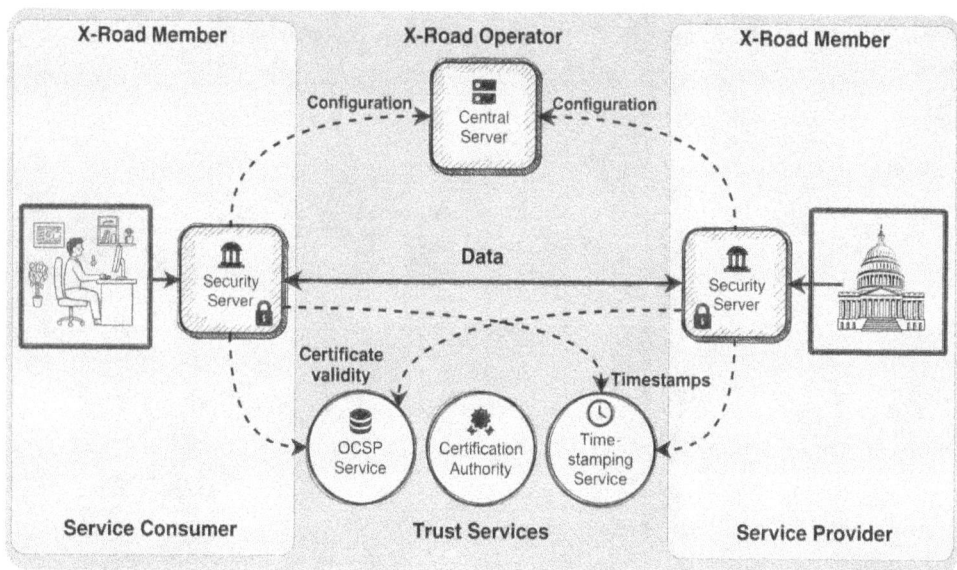

"Instead of creating more layers of bureaucracy, MyVote cuts through them with X-Road—the standardized open source data exchange layer pioneered by Estonia, the leader in digital governance."

"X-Road empowers citizens by connecting them directly to the information and services that are already available, but too often hard to find."

- MyVote Demo Video

The Problem: A Government That Can't Talk to Itself

Imagine trying to have a conversation where every person speaks a different language, refuses to share their notes, and insists you walk to each one individually to get a complete answer. That's exactly how American government systems work today.

The current reality is staggering:

Our federal government operates over 6,000 separate legacy systems that don't communicate with each other.[14] When you apply for benefits, update your address, or try to verify your identity, that information sits trapped in isolated databases. Social Security doesn't automatically share with the IRS. The VA doesn't seamlessly connect with Medicare. Your state DMV has no secure way to verify information with federal agencies.

This isn't just inefficient—it's expensive and dangerous:

- Countless hours are wasted re-entering the same information on multiple government websites, each with different login requirements, security questions, and verification processes.
- Government wastes billions of dollars paying contractors to build custom integrations that break whenever systems are updated.
- Fraudsters exploit the gaps between systems, claiming benefits they are not entitled to because agencies can't quickly verify information across databases.
- Emergency responders lose critical time when disaster strikes because they can't instantly access the data they need from other agencies.
- Your private data becomes more vulnerable as it sits duplicated across dozens of poorly-secured legacy systems.[15]

The tragedy is that the information you need already exists. The services you are entitled to are already funded. They are just locked away behind incompatible systems, outdated technology, and bureaucratic barriers that serve no one.

The Solution: X-Road—The Data Highway America Needs

MyVote integrates X-Road, the proven open-source data exchange layer that transformed Estonia from a post-Soviet state into the world's most advanced digital society. Think of X-Road as a secure, standardized highway system for data—instead of every agency building its own roads (or worse, having no roads at all), everyone uses the same infrastructure with the same rules.

Here's how X-Road could work if it's adopted in the United States:

When you log into MyVote to check your benefits eligibility, you don't manually gather documents from five different agencies. Instead, with your permission, MyVote sends a secure query through X-Road. In milliseconds:

- Social Security confirms your work history
- The IRS verifies your income
- Your state unemployment office checks your status
- Medicare validates your coverage
- The relevant benefit program calculates your eligibility

All of this happens securely, automatically, and with full transparency. You will be able to see exactly which agencies accessed what information, when, and why. Each query is logged, encrypted, and auditable.

The key innovations X-Road brings:

1. Standardization Without Centralization

X-Road doesn't replace existing systems or require agencies to rebuild everything. Instead, it creates a standardized interface layer. Agencies keep their databases, but they allow specific information to pass through X-Road's secure data exchange protocols. It's like ensuring every building has a standard door size—you don't tear down the building, you just make sure people can get in.

2. Security by Design

Every data exchange is encrypted end-to-end. Every query requires authentication. Every access is logged and auditable. Unlike centralized databases that create single points of failure, X-Road distributes data across agencies while maintaining iron-clad security. Even if one system is compromised, the damage is contained.

3. Citizen Control

You decide who accesses your data and when. Need to authorize your accountant to pull your tax transcripts? Grant temporary access through MyVote. Want your doctor to check your Medicare coverage? Approve it with one click. Every permission is time-limited, specific, and revocable.

4. Real-Time Verification

No more waiting weeks for agencies to mail paper confirmations back and forth. When you apply for a mortgage and need to verify your Social Security income, your lender (with your permission) gets instant, verified data directly from the source. Fraud becomes exponentially harder when verification is instant and cryptographically signed.

Proven Results: What Estonia Achieved

Estonia isn't a hypothetical case study—it's a country of 1.3 million people that processes 99% of government services online, saving 844 years of working time annually through digital efficiency.[16] Creating a robust system that continued to operate seamlessly while Russia launched massive cyberattacks.[17]

With X-Road at the core of their digital infrastructure the results are clear:

- **Taxes take 3-5 minutes to file** because the system pre-fills everything from employer reports, banks, and other agencies.

- **Prescriptions are 99% digital**[18]—your doctor enters it once, every pharmacy can see it, and your insurance automatically processes it.

- **Business registration takes 15 minutes**[19] because company data flows automatically between the business registry, tax authority, and statistical office.

- **Medical records follow you seamlessly** between doctors, specialists, and hospitals while you retain complete control over who sees what.

This isn't science fiction. This is Tuesday in Tallinn.

Why This Matters to You

Time: Instead of spending hours navigating government websites and waiting weeks for confirmations, you get instant, accurate information. Filing taxes, applying for benefits, updating your address—all become simple, fast tasks.

Money: Reducing fraud and inefficiency saves taxpayers billions annually. Those savings can lower costs or improve services. For you personally, faster benefits processing means less financial stress during hardship.

Peace of Mind: Your data is more secure, distributed across agency systems with X-Road's encryption and access controls than it is duplicated across vulnerable legacy systems. You gain visibility and control over your information for the first time.

Dignity: You shouldn't have to prove you're a citizen, explain your medical history, or verify your military service over and over to every agency. X-Road means government systems recognize you while respecting your privacy.

The Bottom Line

Integration isn't about building a surveillance state or creating Big Brother. It's about making the government you already fund actually work for you. X-Road has been battle-tested for over 20 years across Estonia, Finland, Iceland, and dozens of other countries. The technology is proven. The security is robust. The benefits are undeniable.

MyVote brings this same empowerment to America. Instead of asking you to trust politicians or bureaucrats it gives you direct access to your government services through systems you can verify, audit, and control.

X-Road eliminates bureaucratic barriers without eliminating privacy protections. It connects government systems without centralizing power. It empowers citizens without exposing data. This is how modern governance works—and MyVote brings it to America.

The information is yours. The services are yours. The government is yours.

It's time it worked like it.

Summary of Best Practices
How MyVote Could Implement
X-Road for America

Phase 1: Federal Foundation

MyVote can begin by connecting core federal services that citizens interact with most: Social Security, IRS, Medicare/Medicaid, Veterans Affairs, and USPS. These agencies handle hundreds of millions of transactions annually and already have digital systems—they just can't talk to each other efficiently.

Phase 2: State Integration

States can also lead the way by creating their own X-Road network, connecting DMVs, unemployment systems, benefit programs, and licensing boards. States that join early gain immediate efficiency benefits and can offer superior services to their citizens.

Phase 3: Expanding the Ecosystem

With citizen permission, private sector entities can verify information through X-Road. Banks can instantly verify income for mortgages. Employers can confirm Social Security numbers. Healthcare providers can check insurance coverage. All with the citizen's explicit, auditable consent.

Critical Safeguards:

- **Open Source Transparency:** *Every line of X-Road code is publicly auditable. Security researchers worldwide can review it, ensuring there are no backdoors or vulnerabilities.*

- **Decentralized Architecture:** *No single database contains all citizen data. Information stays with originating agencies and flows only when needed, with permission.*

- **Privacy by Default:** *You must explicitly authorize each data connection. Your data isn't shared unless you choose to share it.*

- *Audit Trails:* Every query is logged with who requested what, when, and why. You can see your complete data access history.

- *Regular Security Updates:* X-Road benefits from continuous improvement by a global community of developers and security experts.

Chapter 3

Authenticated Users and Digital ID
Restoring Trust in the Digital Town Square

"*Every account is authenticated and secured with industry-leading biometric technology to protect registered citizens from harassment and abuse by anonymous bots and trolls.*"

"*Every voter gets a voice in this secure digital town square, and every local business, political representative, and community organization gets its own authenticated account to help end the fraud and abuse that flood the internet and profit-driven social media.*"

- MyVote Demo Video

The Problem: When Nobody Knows If You're Real

Social media promised to connect us. Instead, it drowned us in fakery.

The crisis we face is unprecedented.

In the 2020 election cycle alone, researchers identified millions of bot accounts spreading political misinformation on Twitter.[20] During the pandemic, fake health accounts convinced millions of Americans to distrust masks and vaccines, trust bleach, and ignore doctors. Right now, as you read this, foreign intelligence services operate thousands of accounts pretending to be American citizens, sowing division and eroding trust in our institutions.

But bots are just the beginning:

- **Troll farms** employ real people working in shifts to harass, intimidate, and silence legitimate voices—especially women, minorities, and anyone who challenges powerful interests.

- **Impersonator accounts** pretend to be your mayor, your congressman, local businesses, or even your neighbors, spreading false information that's impossible to debunk before it spreads.

- **Coordinated inauthentic behavior** creates the illusion of grassroots movements—what looks like thousands of concerned citizens is actually dozens of paid operatives with hundreds of fake accounts.

- **Identity theft** enables bad actors to open accounts in your name, damage your reputation, and even commit fraud while hiding behind your stolen identity.

- **Anonymous harassment** drives good people out of public discourse entirely because there's no accountability when anonymity shields abusers.

The human cost is staggering:

- Teachers quit social media after coordinated harassment campaigns make their lives miserable—and never find out that most of their attackers weren't even real people.
- Local officials receive death threats from accounts they can't report because there is no way to verify who is behind them.
- Small businesses lose customers to fake review accounts they can't challenge.
- Parents make medical decisions based on advice from "doctors" who are actually marketing bots.

And through it all, the platforms profit. Facebook, Twitter, TikTok—their business model depends on engagement, and nothing drives engagement like outrage, fear, and conflict. They have no incentive to solve the bot problem because bots inflate their user numbers, click on ads, and keep real users doom-scrolling through manufactured controversies.

The result? Americans have stopped trusting anything they see online. We've lost faith in our ability to distinguish truth from fiction, real people from fake accounts, genuine grassroots movements from astroturfed manipulation. Our digital town square has become a place where nobody believes anyone and everyone suspects everyone.

This isn't sustainable. Democracy requires informed citizens having good-faith conversations. That's impossible when half the conversation is fake.

The Solution: One Person, One Voice, One Verified Identity

MyVote implements universal authenticated accounts using biometric verification that's already in your pocket. If you have an iPhone or Android device with facial recognition or fingerprint scanning, you have everything you need to prove you are a real, unique human being.

Here's how MyVote authentication works:

Step 1: Initial Registration

When you create your MyVote account, your identity is verified once using a government-issued ID (driver's license, passport, or state ID card) combined with biometric verification. The system confirms:

- You are who you claim to be (matching photo and information).
- You are a registered U.S. citizen or legal resident.
- You are a unique person (not someone who already has an account).

Step 2: Secure Biometric Binding

Your account is then bound to your device's biometric security—Face ID, Touch ID, or Android biometric authentication. This means:

- Only you can access your account using your unique biological characteristics.
- No one can steal your password because your face or fingerprint is your password.
- You can't forget your credentials or write them on a Post-it note.
- Even if someone steals your phone, they cannot access your MyVote account without your biometric data.

Step 3: Multi-Device Security

MyVote is accessible from multiple devices (phone, tablet, computer), but each device requires biometric verification on login. If you get a new phone, you verify your identity again using your existing biometric data plus a secure transfer code. This prevents account takeovers while maintaining convenience.

Step 4: Living Authentication

Modern biometric systems include liveness detection—verifying you are a real person present at the moment, not a photo, video, or deepfake. This means:

- The system cannot be fooled with a picture of you.
- Sophisticated spoofing attempts are detected and blocked.

- You must be physically present to authenticate.

The key difference: Your biometric data never leaves your device. Apple's Secure Enclave and Android's Trusted Execution Environment process biometrics locally and only send encrypted confirmation tokens to MyVote. The platform never stores your fingerprint or face data—it only confirms that your device has been authorized by the user."

Every Voice Authenticated: The Power of Verified Identity

With universal authentication, MyVote creates different account types that restore trust to digital civic engagement.

Verified Citizen Accounts

Every registered voter has a verified personal account with a visual verification badge. When you post, comment, or vote in a poll, everyone knows you are a real person with a verified identity. You can still use a pseudonym for privacy and display any name you choose—but the system guarantees you are a real, unique citizen and not a bot or foreign operative.

Official Representative Accounts

Mayor, city council members, state legislators, and congressional representatives all get official verified accounts that link directly to their position. When they post, you see verification that proves:

- This is really your elected official, not an impersonator.
- They currently hold the office they claim.
- Their account is directly tied to the official government record.

No more fake "Congressman Smith" accounts spreading misinformation. No more wondering if that response from your senator is real or a staffer gone rogue. Official accountability, transparent and verifiable.

Verified Business Accounts

Local businesses, from the corner café to the regional hospital, can get verified business accounts tied to their business license and tax ID. When a restaurant responds to a complaint or a shop announces a sale, you know it's the actual business, not a competitor sabotaging their reputation or a scammer phishing for credit cards.

Authenticated Organization Accounts

Nonprofits, community groups, churches, schools, and civic organizations will verify their accounts using their EIN and official registration information. When you read a comment by "Springfield Parents for Better Schools," you can verify it's a real registered organization, see who the officers are, and confirm they are actually from Springfield—not an astroturf group funded by out-of-state interests pretending to be local.

Professional Credential Verification

Doctors, lawyers, teachers, and other licensed professionals can link their professional credentials to their accounts. When someone giving medical advice claims to be a doctor, you can verify they are actually licensed and in good standing—not a wellness influencer selling snake oil.

Why Anonymity's Time Is Over (In Civic Spaces)

Some critics will cry, "But what about anonymity? What about whistleblowers?"

Let's be clear: MyVote is for civic engagement, not journalism. There are appropriate places for anonymity—investigative journalism, whistleblower protections, support groups for abuse survivors. MyVote is not trying to replace those spaces.

MyVote is the digital equivalent of a town hall meeting. In a real town hall, you can see who's speaking. You know if the person asking a question is actually your neighbor or a paid operative from out of town. You can recognize when the same person keeps coming to the microphone pretending to represent different viewpoints. That's not surveillance—it's transparency.

The benefits of authenticated civic spaces far outweigh the costs.

1. Accountability Restores Civility

When someone can't hide behind anonymity, they might think twice before posting. Research shows that authenticated online spaces and identity-verified settings reduce aggressive behavior and foster greater accountability.[21] Not because of censorship—but because accountability encourages the same social norms that make in-person community meetings functional.

2. Stopping Foreign Interference

Russia, China, and Iran have spent millions creating fake American personas to influence our elections and divide us.[22] Authentication makes this exponentially harder. They can try to create all the fake accounts they want —they will never be able to pass biometric verification tied to real U.S. identity documents.

3. Ending Bot Manipulation

Bots don't have fingerprints. Bots can't pass facial recognition. One authenticated human = one account. Period. This destroys the power of well-funded troll farms, astroturf campaigns, and anyone trying to manufacture fake consensus.

4. Protecting the Vulnerable

Paradoxically, authentication protects privacy better than anonymity. Right now, anyone can impersonate you or stalk you across unverified platforms. With authentication, harassment becomes traceable and prosecutable. Identity theft becomes nearly impossible.

5. Rebuilding Trust

When we know every person in a conversation is real, when we can verify that our elected officials are actually who they claim, when we can trust that the local business page is actually local—suddenly, productive conversation becomes possible again.

Real-World Results: What Authentication Achieves

Countries and platforms that have implemented strong authentication have seen transformative results.

South Korea's Real-Name System (2005-2012) effectively reduced anti-social behavior by up to 30% at the national level.[23] though it was ultimately repealed due to data breaches and constitutional concerns—problems that MyVote's decentralized biometric approach solves. The Korean Constitutional Court found that the real-name policy reduced malicious comments from 13.9% to 13.0% (a decrease of only 0.9 percentage points),[24] demonstrating that traditional identity verification without strong security is insufficient. MyVote builds on both the successes (behavioral changes) and failures (centralized data vulnerability) of South Korea's experiment.

Estonia's Digital ID enables 99% digital government services while maintaining extremely low fraud rates and high citizen satisfaction. Authentication hasn't destroyed privacy—it has enhanced it by giving citizens control.

Facebook's Real-Name Experiment showed that even weak authentication (just asking for real names) reduced harassment and improved conversation quality. Imagine what strong biometric verification achieves.

China's Social Credit System is often cited as the nightmare scenario—but it's the perfect example of what MyVote is NOT. China's system is government surveillance tracking every purchase, conversation, and movement to assign obedience scores. MyVote is a publicly funded and auditable authentication system designed solely for civic participation. MyVote identity doesn't follow a citizen to Amazon, track browsing, or monitor private messages. It exists solely to verify that only real citizens are participating in democratic processes.

Why This Matters to You

Safety: MyVote allows participation in civic conversations without fear of anonymous harassment campaigns.[25] When someone crosses the line, they can be held accountable. When someone is threatening, law enforcement can investigate instead of hitting a wall of anonymity.

Trust: You can believe that the person presenting themselves as your mayor actually is your mayor. Community organizations are verified as legitimate. And fake astroturf campaigns can't pretend to be real grassroots movements.

Empowerment: Your voice carries weight because everyone knows you're a real person, not a bot inflating engagement metrics. One person, one voice —finally meaningful in the digital age.

Privacy: Paradoxically, strong authentication protects privacy better than anonymity. Nobody can impersonate you, steal your identity, or frame you for things you didn't say. You control your verified identity instead of hoping nobody abuses the anonymous chaos.

The Bottom Line

One authenticated human, one verified voice. No bots. No trolls. No foreign interference. No anonymous harassment. Just real citizens engaging in real democracy, protected by biometric security you already use every day—now applied to the civic space where your voice matters most.

The technology exists. The need is urgent. The choice is ours.

Real people. Real conversations. Real democracy.

That's the promise of authenticated civic engagement. That's the foundation MyVote builds upon.

Summary of Best Practices
How MyVote Implements
Authentication Securely

Privacy by Design

- _Your biometric data never leaves your device._
- _MyVote stores only encrypted tokens, not actual fingerprints or facial scans._
- _You choose what name to display publicly (real name or pseudonym)._
- _Your full identity verification is visible only to you and system administrators investigating abuse._
- _Your participation in discussions can use privacy-preserving display names while maintaining back-end accountability._

Progressive Verification Levels

Not all interactions require the same verification:

- **_Viewing public information:_** _No account needed_
- **_Participating in discussions:_** _Verified citizen account_
- **_Voting in binding referendums:_** _Full identity + voter registration verification_
- **_Commenting on sensitive topics:_** _Additional verification checks may be required_

Accessibility Accommodations

- _Alternative verification methods for citizens who cannot use standard biometrics (injuries, disabilities, etc.)._
- _Assisted verification at public libraries and government offices for those without smartphones or computer access._
- _Clear, simple processes with multilingual support and assistance._

Regular Security Audits

- *Third-party penetration testing quarterly*
- *Open-source verification code for public review*
- *Bug bounty program rewarding security researchers who find vulnerabilities*
- *Transparent reporting of any breaches or attempted attacks*

Legal Safeguards

- *Strong warrant requirements for any law enforcement access to identity data*
- *Automatic notification when your data is accessed (except in specific court-approved emergency circumstances)*
- *Regular transparency reports showing aggregate statistics on verification, access requests, and security incidents*

Chapter 4

Representation and Accountability
Your Representatives, Finally Within Reach

"For too long, even the most basic access to our elected representatives has been dictated by money and sensational media coverage. But representation, accountability, and transparency are the cornerstones of e-governance."

"With MyVote, your local, state, and federal representative offices are just a text chat away, to help you with everything from a pothole on your street to Social Security and Medicare, and every interaction within the MyVote system is logged and timestamped as part of the official public record."

- MyVote Demo Video

The Problem: A Representative Democracy Where Nobody Can Reach Their Representatives

Democracy's most fundamental promise is simple: you elect someone to represent your interests, and they work for you. But somewhere between the Founding Fathers and today, that promise became a cruel joke.

The access crisis is real and getting worse. We might call or email our congressperson, but how do we know they're listening or doing something about our concerns or voting in our best interests?

Try calling your congressman's office right now. If you're lucky enough to get through the busy signal, you'll reach a 22-year-old intern who takes a message that may or may not ever reach the person you elected. Want a response? The average congressional office receives tens of thousands of constituent contacts annually,[26] with some offices receiving over 81 million messages collectively in 2022.[27] Congressional offices typically have staff to meaningfully respond to a fraction of these contacts. Your odds of getting actual help are roughly the same as winning a decent prize in the lottery.

Here's what representation looks like in America today...

Access Goes to the Highest Bidder: Lobbyists spend over $4 billion annually[28] to ensure their calls get answered, their meetings get scheduled, and their concerns get addressed. They don't wait on hold. They don't fill out web forms that disappear into the void. They text the representative's personal cell phone and get lunch scheduled by Friday.

Media Coverage Determines Priority: Your representative spends more time preparing for cable news appearances than reading constituent mail. A viral tweet gets more attention than a thousand thoughtful letters. Outrage drives response, not substance. The squeaky wheel gets the grease—but only if the squeaking happens on Twitter, where it might trend.

Geographic Barriers Lock Out Rural America: If you live in rural Montana, your representative's office might be 300 miles away. Town halls happen in population centers, leaving vast swaths of constituents with no practical way to attend. Phone lines are overwhelmed. In-person visits require taking time off work and driving for hours. Your representative might as well be on another planet.

The Casework Crisis: When you need help navigating the federal bureaucracy—Social Security isn't processing your disability claim, the VA lost your medical records, Medicare denied coverage for a necessary procedure—your representative's office is supposed to help. That's what casework is. But most offices are so overwhelmed that cases sit for months. Members often receive over a thousand constituent requests per year for casework assistance.[29] By the time they get to yours, you've already lost your home, missed critical medical treatment, or given up entirely.

Zero Transparency: When your representative meets with constituents, lobbyists, or donors, there is usually no detailed, real-time public record of those meetings. You don't know who is getting their ear, how often, or on which specific issues. Floor votes are public, and committee hearings are announced, but the thousands of private conversations and behind-the-scenes decisions that shape those votes remain largely opaque. You elected them, but you still have almost no clear picture of how they actually spend their time or who has the most access.

The Broken Feedback Loop: Even when you do somehow manage to get a response, there's no accountability mechanism. Your representative's office might send you a form letter that doesn't address your actual concern, or promise to "look into it" but then nothing happens. Or they could take a position in their response to you, then vote the opposite way—and there's no record connecting their private promise to their public betrayal.

The human cost compounds daily:

- A disabled veteran waits 18 months for VA benefits because the congressional caseworker never followed up.
- A small business owner can't get answers about contradictory EPA and state environmental regulations, so they just shut down rather than risk massive fines.
- A grandmother loses her Social Security survivor benefits due to a bureaucratic error and can't get anyone to fix it—she exhausts her savings and loses her home while waiting for help that never comes.

Meanwhile, the defense contractor who wants the representative to support their latest $500 million procurement request? They get a meeting within 48 hours. The pharmaceutical lobbyist concerned about drug pricing legislation? Their call is returned before lunch. The hedge fund manager who maxed out campaign contributions? Coffee next Tuesday work for you?

This isn't representation. This is access-for-sale masquerading as democracy.

The Solution: Direct Digital Access to Every Elected Official

MyVote creates direct, documented, democratic access to every elected official from your city council member to the President of the United States. Not through gatekeepers. Not through donation requirements. Not through media spectacles. Through a simple, transparent, accountable system that treats every constituent equally.

Here's how MyVote transforms representation…

Direct Messaging to Representative Offices

Every elected official—local, state, and federal—has an official MyVote account staffed by their office. You can send a direct message to your:

- **City Council Member:** Report that pothole on Elm Street, ask about the proposed zoning change, inquire about local parks funding.
- **Mayor:** Question budget priorities, request help with a city department, propose a community initiative.
- **State Representative:** Get help with state benefits, ask about education funding, voice concerns on pending legislation.
- **State Senator:** Request assistance with professional licensing, discuss infrastructure projects, advocate for policy changes.
- **Governor:** Escalate unresolved state agency issues, ask about executive orders, discuss statewide priorities.
- **U.S. House Representative:** Get help with federal agencies (Social Security, VA, IRS, Medicare), discuss federal legislation, request assistance with immigration cases.

- **U.S. Senators:** Same as House plus judicial nominations, treaties, and Senate-specific issues.
- **President:** While the President's office handles millions of contacts, MyVote creates a direct, documented channel for urgent constituent needs.

Every message is:

- Delivered directly to the representative's official MyVote office account (no getting lost in email spam filters).
- Timestamped and logged as part of the permanent public record.
- Guaranteed a response within publicly posted timeframes (offices set their own standards, and MyVote timestamps track their response).
- Tracked to completion for casework requests, with status updates visible to you and recorded publicly.

The Casework Revolution

Casework—helping constituents navigate government bureaucracy—is where MyVote creates the most immediate impact. Here's how it works…

1. You Submit a Request

Example: "My Social Security disability claim has been pending for 14 months with no response. Case number: XXX-XX-XXXX. I've called the local office six times. I'm about to lose my apartment. Please help."

2. Automatic Case Assignment

Your message is immediately assigned to a caseworker in your representative's office. You receive a case number and acknowledgment within 24 hours (or whatever standard that office has publicly committed to).

3. MyVote System Integration

Because MyVote integrates with X-Road, the caseworker can (with your explicit permission) instantly access the relevant information:

- Social Security claim status directly from SSA systems.

- Processing timelines and current backlog data.
- Case history and documentation.
- Contact information for the specific SSA employee handling your case.

4. Documented Intervention

The caseworker contacts SSA on your behalf—and every step is logged:

- "10/26/25 10:30 a.m. - Initial contact with SSA Regional Office."
- "10/26/25 2:45 p.m. - Spoke with Claims Examiner Rodriguez, case in medical review queue."
- "10/27/25 11:15 a.m. - Requested expedited review due to housing instability."
- "10/29/25 3:20 p.m. - Medical review completed, claim approved, payment processed."

5. Transparent Updates

Every update is visible in real-time through MyVote. No more calling and leaving voicemails. No more wondering if anyone is actually working on your case. Complete transparency from start to finish.

6. Public Accountability

The aggregate data becomes part of the public record:

- Representative Smith's office completed 847 casework requests this quarter.
- Average time to resolution: 12 days.
- 94% constituent satisfaction rating.
- Most common issues: Social Security (38%), Veterans Affairs (24%), Medicare (18%).

This isn't some glossy brochure statistic—it's real data from real cases, auditable and transparent.

Office Hours Go Virtual and Inclusive

Representatives can host virtual office hours through MyVote:

- Text-based town halls where constituents ask questions and representatives respond in real-time, all transcribed and archived.
- Audio/video sessions for more complex discussions, recorded and posted publicly.
- Topic-specific forums (e.g., "Small Business Owners Discussion" or "Veterans Services Q&A").
- Multilingual support ensuring non-English speakers have equal access.
- Accessibility accommodations including screen reader compatibility and captioning.

No more driving three hours to be locked out of a town hall that only seats 50 people. No more elderly or disabled constituents unable to participate because the venue isn't accessible. No more working parents missing events because they're scheduled at 2 p.m. on a Tuesday.

Virtual office hours through MyVote mean every constituent can participate regardless of location, mobility, work schedule, or childcare constraints.

Legislative Transparency in Real-Time

Every official action is documented.

Voting Records: Not just how they voted, but:

- The full text of what they voted on (automatically pulled from official sources).
- An explanation for their vote (if provided).
- How that vote aligns with campaign promises (MyVote tracks these).
- How their vote aligns with constituent sentiment (based on MyVote polling in their district).

Meeting Logs: Public officials using MyVote for scheduling create automatic transparency:

- "10/26/25 - Met with Springfield Teachers Union - discussed education funding."
- "10/27/25 - Meeting with TechCorp lobbyist - 5G infrastructure legislation."
- "10/28/25 - Virtual office hours with District 12 constituents - 43 participants."

This isn't surveillance of private life—it's transparency about official duties. Who they meet with in their official capacity, what they discuss, and how they spend time on the public payroll.

Response Time Tracking:

- Average time to respond to constituent messages.
- Percentage of messages that receive substantive responses vs. form letters.
- Casework completion rates and average resolution times.
- Comparison metrics showing how they rank against other representatives.

Campaign Promise Tracking: MyVote archives campaign promises and tracks them against actual performance:

- "Promised to vote against tax increases" → Voted YES on HB 2847 (property tax increase).
- "Promised to hold monthly town halls" → Held 3 town halls in 12 months.
- "Promised to make veterans' issues a priority" → Voted NO on veterans' healthcare funding increase.

This isn't editorializing—it's just facts. Constituents can decide for themselves whether campaign promises match governing reality.

Real-World Results: What Digital Representation Achieves

Estonia's e-Democracy Experience:

- 99% of government services available online 24/7.
- Citizens directly petition parliament through digital platforms— 1,000 signatures trigger mandatory parliamentary review.[30]
- Every government decision is documented and accessible to citizens.
- Politicians' meeting schedules are public and searchable.

- **Result:** Estonia ranks #3 globally in government transparency and #1 in digital governance.

Iceland's Better Reykjavík Platform:

- Digital platform where citizens propose ideas and city officials respond publicly.
- Over 700 citizen proposals implemented.[31]
- City officials must respond to popular proposals with concrete action plans.
- Over 70,000 people have participated out of a population of 120,000.[32]
- **Result:** 12.5% regular participation rate in participatory budgeting (vs. traditional 5-7% town hall attendance), massively increased civic engagement.[33]

Taiwan's vTaiwan Digital Democracy:

- Digital platform for policy consultation where citizens and government collaborate.
- Used to resolve complex issues like Uber regulation, online alcohol sales, and digital currencies.
- Requires government officials to respond to citizen input with documented reasoning.
- **Result:** 80%+ approval ratings for policies developed through the platform, vs. 30% for traditional top-down legislation.

Madrid's Decide Madrid Platform:

- Citizens propose and vote on municipal projects.
- City government must implement winning proposals.
- All official meetings and decisions are documented publicly.
- **Result:** 390,000+ registered users, €100 million in citizen-directed spending, dramatically improved trust in local government.

Why This Matters to You

Access: You're no longer shut out by gatekeepers, geography, or your inability to write a $5,000 check. Your representative is as accessible as your email, and they're required to respond.

Action: When you need help—with Social Security, veterans benefits, Medicare, or any government service—you get real assistance instead of voicemails and form letters. Your problems get solved, not ignored.

Accountability: You can see exactly how your representative spends their time, who influences them, and whether their actions match their promises. No more voting blind. No more believing campaign rhetoric that evaporates the day after the election.

Empowerment: Your voice matters equally to every other constituent. Not because politicians suddenly became altruistic, but because MyVote creates structural accountability. They can't ignore you when their response times, constituent satisfaction ratings, and casework performance are public record.

Community: You can see what issues other constituents are raising, how the representative responds to them, and whether there's a pattern of responsiveness or neglect. Collective knowledge becomes collective power.

The Bottom Line

Representative democracy is broken, but not beyond repair. Technology helped create the problem—giving lobbyists instant access, rewarding media spectacle, and making it easier to ignore constituents than engage them. Now technology can fix it. MyVote does not demand that representatives work harder; it demands that they work transparently: building accountable systems for constituent service, acknowledging every voice, responding with clear reasoning, and leaving a public trail of every interaction, decision, and promise. Direct access to every elected official, documented responses to every constituent, and transparent records of every official action—logged, timestamped, and permanently public, turning "your representative works for you" from a slogan into a measurable reality for the first time in generations.

Summary of Best Practices
How MyVote Implements
Accountable Representation

Tiered Response System

Not every message requires the representative's personal attention. MyVote implements smart prioritization:

- ***Tier 1 - Automated Information:*** *"Where do I vote?" "What district am I in?" "When is the next election?"* → *Instant automated responses with accurate, verified information.*
- ***Tier 2 - Staff Response:*** *Standard casework, constituent questions, policy inquiries* → *Handled by trained office staff with documented response requirements.*
- ***Tier 3 - Senior Staff/Representative:*** *Complex cases, urgent situations, policy discussions requiring representative input* → *Escalated appropriately with priority handling.*

The key difference: Every tier is documented and accountable. Even automated responses are logged. Every staff reply is timestamped and attributed. Every escalation is tracked.

Constituent Privacy Protection

While interactions are logged for accountability, personal information is protected:

- *Public records show aggregate data, not individual constituent details.*
- *Specific casework is visible to you and the representative's office, not broadcast publicly.*
- *Public forums and town halls are fully public (you're speaking in a public space).*
- *Private constituent communications remain private but are logged as "Contact received/responded to" without content details.*

- **You control what's public:** *You can choose to make your interactions public if you want others to see how responsive (or unresponsive) your representative is.*

Office Accountability Standards

Each representative's office publicly commits to standards:

- *"We respond to all messages within 48 business hours."*
- *"We provide substantive responses, not just form letters."*
- *"Casework requests receive status updates every 5 business days until resolved."*
- *"We hold virtual office hours twice monthly."*
- *"We respond to MyVote survey questions about pending legislation."*

MyVote then tracks performance against these commitments. Did they actually respond within 48 hours? Are constituents satisfied with the quality of responses? Are casework promises being kept?

Performance Dashboards

Every representative has a public performance dashboard showing:

- **Response metrics:** *Average response time, response rate, constituent satisfaction scores.*
- **Casework performance:** *Cases opened, cases closed, average resolution time, success rate.*
- **Accessibility:** *Office hours held, constituent meetings conducted, town halls hosted.*
- **Voting alignment:** *How their votes align with district sentiment (based on MyVote constituent polling).*
- **Transparency score:** *Percentage of meetings logged, campaign promises tracked, public engagement level.*

These aren't subjective ratings—they're objective metrics derived from MyVote system data.

Constituent Feedback Loop

After every interaction, you can rate the experience…

- *"Did this resolve your issue?"*
- *"Was the response timely?"*
- *"Did you receive a substantive answer or a form letter?"*
- *"Would you recommend this office to other constituents?"*

Aggregate feedback is public. Bad representatives can't hide behind slick PR campaigns when their MyVote constituent satisfaction rating is 2.1 stars.

Chapter 5

Personalized Dashboard
Your Command Center for Civic Engagement

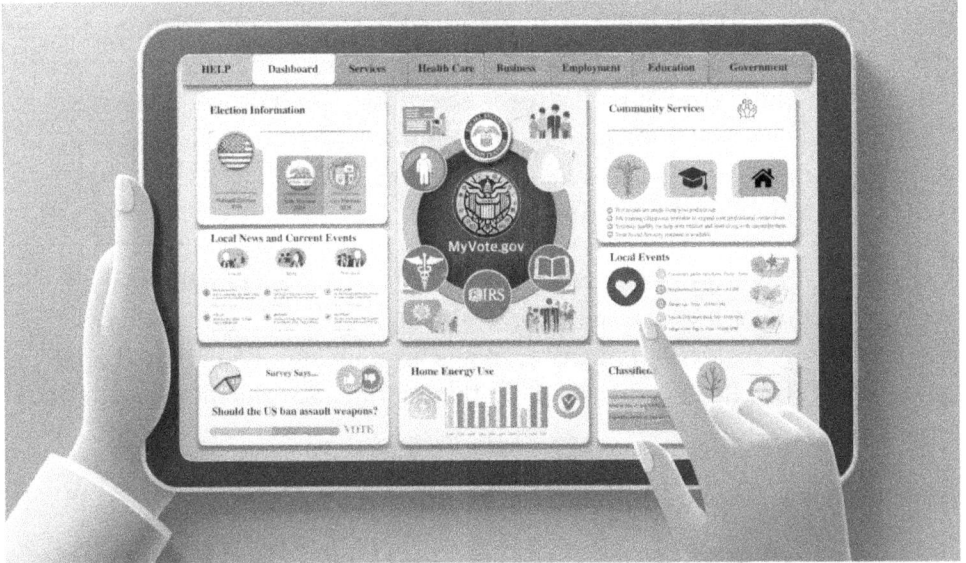

"Every citizen gets a personalized dashboard that will keep you informed about the issues that concern you the most. You can follow the topics, community leaders, and elected representatives that interest you while sharing the news, events, and opinion polls that will help your community stay informed and engaged."

"We're also leading the battle against misinformation by labeling fake news and AI-created content and providing links to all the facts. We know knowledge is power. So we're making MyVote your gateway to public records and educational resources like libraries, historical records, and government archives that will allow us to build a better future by learning from our collective past."

- MyVote Demo Video

The Problem: Drowning in Information, Starving for Knowledge

We all want to be an informed citizens, but the system seems designed to make that impossible.

The information overload crisis is breaking civic engagement:

Every day, we are bombarded with thousands of messages screaming for our attention. Our phone buzzes with breaking news alerts that contradict each other within hours. Social media feeds overflow with outrage, conspiracy theories, and emotional manipulation disguised as news. Cable news networks turn every issue into a five-alarm fire designed to keep everyone watching through the next commercial break.

Meanwhile, the information we actually need—the city council meeting agenda, the text of legislation our representative is voting on tomorrow, the real facts about that local issue everyone's arguing about—is buried on a government website designed in 2003 that requires seventeen clicks to find anything useful.

Here's what trying to be an informed citizen looks like today...

Information Scattered Everywhere: Want to know what your city council is doing? Check their website (maybe). What about your school board? Different website. State legislature? Another site. Your congressional representative? Yet another platform. Federal agencies? Dozens of different portals. Nothing connects. Nothing centralizes. You need a PhD in website navigation just to find basic information.

Critical Information Hidden: The most important decisions affecting your life happen in obscure meetings with barely any notice. A zoning change that will put a factory next to your child's school? Mentioned once in a legal notice in the local paper and on a government website you've never heard of. By the time you find out, the decision's been made and the public comment period is closed.

Misinformation Floods the Zone: For every legitimate news article, there are fifty misleading posts, manipulated videos, and outright fabrications. Russia, China, and Iran spend millions creating sophisticated propaganda.

Domestic political operatives create fake "news" sites that look legitimate. AI-generated content is indistinguishable from real journalism. It's literally impossible to tell what's true anymore.

Algorithm Manipulation: Social media platforms don't show what's important—they show what keeps you engaged. Facebook's algorithm promotes outrage because angry people click more ads.[34] TikTok's algorithm optimizes for watch time, not truth. YouTube's recommendations can lead people toward more extreme content.[35] We are not getting informed —we are being manipulated.

No Context, All Emotion: A bill gets one-sentence headlines: "Congress Votes to Slash Medicare!" What actually happened? A 0.3% reduction in the rate of growth over ten years, maintaining current benefit levels. But nuance doesn't drive clicks, so we never get the full story. We just get rage bait.

Fake Grassroots Everywhere: That "concerned parents group" flooding our feed with school board posts? Funded by a national political organization. That "local business coalition" opposing regulations? A shell group for a multinational corporation. That "community petition" with 10,000 signatures? Most are bots and fake accounts. We can't tell what's real grassroots concern and what's astroturfed manipulation.

Educational Resources Siloed and Inaccessible: Want to understand the historical context of an issue? Good luck finding it. Libraries have archives, but they're not digitized. Government records exist, but they're scattered across agencies. Historical documents are locked in physical repositories. Academic research is behind paywalls. The knowledge exists—we just can't find it when we need it.

Cognitive Exhaustion: We are so overwhelmed trying to separate truth from lies, so exhausted from information overload, so frustrated by the impossibility of staying informed, that we just give up. It's easier to tune out entirely or retreat into whatever information bubble confirms our existing beliefs.

The human cost of information chaos...

A grandmother believes false information about vaccines and convinces her daughter not to immunize her grandchildren. A small business owner votes against his own economic interests because he's been fed sophisticated misinformation about tax policy. A community opposes a needed infrastructure project based on fabricated public safety concerns spread by profit-driven competitors. A parent pulls their child from public school based on viral lies about curriculum that was never actually taught.

Meanwhile, corporations and special interests have entire teams monitoring legislative activity, tracking regulatory changes, and analyzing policy impacts. They are informed. They are engaged. They are effective. All of us are just struggling to figure out what's true.

This isn't an accident. An uninformed, confused, exhausted citizenry is easier to manipulate and control. And information chaos serves powerful interests while disempowering you.

The Solution: Your Personalized Civic Command Center

MyVote creates a single, personalized, verified information hub that cuts through the noise and delivers exactly what you need to be an informed, engaged citizen—nothing more, nothing less.

Here's how your MyVote Dashboard works...

Intelligent Personalization: Follow What Matters to You

When you set up MyVote, you customize your dashboard based on your actual interests and responsibilities.

Location and Jurisdictions

The system automatically identifies all governing bodies that affect you:

- Springfield City Council (Ward 3)
- Springfield Public Schools Board
- Sangamon County Board of Supervisors
- Illinois State Legislature (District 87 House, District 44 Senate)
- U.S. House (IL-13)

- U.S. Senate (Illinois)
- Relevant federal agencies (based on your interests and demographics)

Topics of Interest

You choose what issues you want to track:

- Education policy (you have school-age children)
- Small business regulations (you own a bakery)
- Veterans affairs (you're a veteran)
- Environmental policy (you care about climate)
- Healthcare policy (your mother is on Medicare)

Representatives

Follow the elected officials who represent you:

- Mayor
- City Council Members
- State Representatives
- State Senator
- U.S. House Representative
- U.S. Senators

Community Leaders and Organizations

Follow verified accounts for:

- The Chamber of Commerce
- Local Veterans Organization
- School principals
- Environmental advocacy groups
- Local newspaper

Every morning, you open your Dashboard and immediately see...

Today's Relevant Activity:

- "City Council meets tonight at 7 p.m. - Agenda includes zoning proposal affecting Main Street businesses."

- "State Rep. Williams voting today on HB 2847 (Small Business Tax Relief Act) - District polling shows 73% support."
- "New study published on veteran healthcare wait times in your region."
- "Springfield School Board election candidate forum this Thursday."

Upcoming Decisions That Affect You:

- "Public comment period closes Friday for EPA regulation affecting small bakeries - Submit your input."
- "Town hall on proposed highway expansion through Springfield - Tuesday 6 p.m. at Community Center."
- "State budget proposal includes $50M cut to education funding - Contact your state rep."

Personalized News and Analysis:

Not random news—news specifically relevant to your chosen topics and location:

- Local journalism about Springfield issues
- Verified national reporting on education policy
- Veterans Affairs policy updates
- Environmental policy analysis
- Healthcare reform developments

All verified, contextualized, and directly relevant to your life.

Real-Time Legislative Tracking: Never Miss What Matters

Your dashboard tracks every bill, regulation, and policy development relevant to your interests, showing you legislation at every level.

Bill Tracking:

- **Local:** "Ordinance 2025-47: Plastic Bag Ban - Planning Commission recommends approval - City Council vote scheduled 11/15."
- **State:** "HB 2845: Small Business Tax Relief - Passed House 89-28 - Now in Senate Finance Committee."

- **Federal:** "HR 1234: Veterans Healthcare Expansion Act - House vote expected next week - Your rep hasn't declared position yet."

For Each Bill, You Get:

- Plain-language summary (not legalese—actual English explaining what it does)
- Full text (if you want to read the actual legislation)
- Fiscal impact (what it costs, who pays, what it funds)
- Stakeholder positions (who supports, who opposes, why)
- Constituent polling (what people in your district think)
- Your representative's position (and how it aligns with district opinion)
- Action options (contact your rep, sign petition, share with others)

Automatic Alerts:

- "HB 2847 scheduled for vote tomorrow - your rep hasn't declared a position - contact them now."
- "City Council added a surprise item to tonight's agenda: rezoning proposal affecting your neighborhood."
- "Final day to submit public comment on small bakery regulations."

Historical Context:

- "Similar legislation failed in 2018 due to lobbying by [Industry Group]."
- "This is the third attempt to pass this reform in five years."
- "Previous version was vetoed by the Governor - key differences in this version."

Verified News and Information: Truth Over Noise

MyVote aggregates news from verified sources and clearly labels everything.

Source Verification:

Every news item shows:

- **Verified Journalism** - From credentialed news organizations with editorial standards.
- **Official Government Information** - Direct from government agencies.
- **Expert Analysis** - From verified subject matter experts with credentials.
- **Opinion/Commentary** - Clearly labeled as opinion, not fact reporting.
- **User-Generated Content and Comments** - From authenticated citizens, clearly distinguished from journalism.

AI Content Labeling:

MyVote technology will also automatically identify and label:

- **AI-Generated Text** - Content written by AI systems.
- **AI-Manipulated Images** - Photos altered by AI (deepfakes, manipulations).
- **AI-Generated Video** - Synthetic video content.

This doesn't ban AI content—it just ensures you know what you are consuming.

Misinformation Flagging:

Content flagged by fact-checkers shows:

- **False** ✖ - Definitively untrue, fact-checked by multiple verified sources.
- **Misleading** ! - Technically true but missing critical context.
- **Unverified** ? - Claims that cannot currently be confirmed.
- **Context Needed** ℹ️ - True but requires additional information to understand.

Each Flag Includes:

- Links to fact-checks from multiple independent sources.
- Explanation of what's false/misleading and why.
- Links to accurate information on the topic.
- Source credibility ratings.

Example:

You see a viral video in your feed: "Springfield Mayor caught on camera saying she'll raise property taxes 40%!"

MyVote labels it:

- **AI-Manipulated Video** - Audio has been altered using AI voice synthesis.
- **Misleading** - Quote taken out of context from a longer discussion.
- Fact-checks from 3 sources provided.
- Link to full, unedited video of the meeting.
- Mayor's verified MyVote account with her own response.

You can still watch it if you want—but you know exactly what you're looking at.

Direct Access to Public Records and Archives

MyVote is your gateway to the knowledge that's been locked away.

Integrated Government Records:

- Meeting minutes and agendas for every government body that represents you.
- Voting records for every elected official, searchable and downloadable.
- Budget documents in both full detail and plain-language summaries.
- Public records requests simplified—file and track FOI requests directly through MyVote.
- Regulatory filings affecting your business or industry.
- Court records for public cases (properly anonymized for privacy where required).
- Property records searchable by address.

- Business licenses and permits - see who's operating what in your area.

Digital Library Integration:

MyVote connects you to:

- Library of Congress digital collections - millions of historical documents, photos, recordings.
- National Archives - founding documents, historical records, declassified materials.
- State archives - local history, legislative history, historical photos and documents.
- University research repositories - academic studies and papers on policy issues.
- Public library systems - ebooks, research databases, local history collections.

Searchable by Topic:

Researching education policy? MyVote finds:

- Historical education legislation and outcomes.
- Academic research on education reforms.
- Government reports on education spending and performance.
- Primary source documents from education debates throughout history.
- Case studies from other states and countries.

All in one search, with proper context and verification.

Educational Context for Current Events:

When you read about a current issue, MyVote provides historical context:

Reading about a border security debate? See links to relevant information about:

- Previous immigration legislation and its outcomes.
- Historical immigration patterns and policies.
- Academic research on immigration's economic impacts.
- Primary sources from past immigration debates.

- International comparisons showing how other countries handle similar issues.

Not to tell you what to think—but to give you the knowledge to think critically.

Community Engagement Hub

Your dashboard isn't just for consuming information—it's for engaging with your community in a virtual town square.

Local Events Calendar:

- Town halls and public meetings
- Candidate forums and debates
- Community organizing events
- Educational workshops on civic topics
- Voter registration drives
- Public comment opportunities

Discussion Forums by Topic:

- Authenticated citizens discussing issues (no bots, no trolls)
- Moderated to prevent abuse while encouraging robust debate
- Verified experts available to answer questions
- Representatives can participate directly
- Community consensus-building on local issues

Sharing Tools:

- Share verified news articles with friends (not misinformation)
- Forward meeting notices to neighbors
- Organize carpools to town halls
- Create study groups for ballot initiatives
- Build coalitions around issues you care about

Polls and Surveys:

- Quick pulse checks on emerging issues
- In-depth opinion surveys on complex topics
- See how your community thinks about issues

- Compare your views to neighbors'
- Share results with representatives

Notification Control Center

You control exactly how and when MyVote reaches you:

Notification Priorities:

- **Urgent:** Immediate votes, last-minute meeting changes, breaking news directly affecting you.
- **Important:** Upcoming votes, deadlines, major policy developments.
- **Informational:** General updates, news articles on topics you follow, community events.
- **Digest:** Weekly summary of activity (for people who don't want constant notifications).

Channel Preferences:

- Push notifications to phone
- Email summaries
- Text message alerts (for urgent items only)
- In-app notifications

Granular Control:

- "Notify me immediately about Springfield City Council activity."
- "Weekly digest for state legislature activity."
- "Urgent only for federal legislation."
- "Notify me about veterans issues but not other topics."

You decide. MyVote adapts to your needs, not the other way around.

Real-World Results: What Information Democracy Achieves

Estonia, Finland, Taiwan, and Denmark are just 4 the 25 nations with functional electronic governance systems that prove everyday how successful digital democracy can be.

Estonia's Public Information Architecture:

- All government data is centralized and accessible through a single platform.
- Citizens can see every government decision and action affecting them.
- **Result:** 83% satisfaction with administrative services (2023), high civic engagement.[36]

Finland's Media Literacy Success:

- Comprehensive media literacy education integrated into daily life.
- Citizens trained to recognize misinformation and verify sources.
- **Result:** Ranked #1 globally in resistance to misinformation for five consecutive years (2017-2023), despite being targeted by Russian disinformation campaigns.[37]

Taiwan's Civic Tech Movement:

- g0v (gov-zero) platform aggregates government data and makes it accessible.
- Citizens create tools to visualize and understand public information.
- **Result:** Highly informed citizenry that successfully resists Chinese disinformation operations.[38]

Denmark's Digital Government Integration:

- Single digital portal for all government services and information.
- Personalized dashboard showing relevant government activity.
- **Result:** 90%+ digital government usage, extremely high citizen satisfaction and trust.[39]

Why This Matters to You

Time: Instead of spending hours hunting for information across dozens of websites, you get everything relevant in one place. Five minutes on MyVote gives you more useful civic information than an hour of doomscrolling social media.

Truth: You know what's real and what's not. AI labels, fact-checks, and source verification mean you can trust what you're reading—and you know when you shouldn't.

Power: Knowledge really is power. When you understand what's happening, why it matters, and what you can do about it, you become a more effective citizen. You make better decisions. You engage more productively.

Connection: You discover you're not alone. Your neighbors care about the same issues. Your community is more aligned than mass-media makes it seem. You find common ground and build coalitions.

Sanity: The information overload stops. The emotional manipulation ends. The algorithm chaos disappears. You get what you need without the exhaustion.

The Bottom Line

For too long, powerful interests have benefited from keeping citizens confused, overwhelmed, and misinformed—because a confused citizen can be manipulated, an exhausted citizen gives up, and a misinformed citizen makes bad decisions. MyVote changes that equation by giving you a personalized, verified, comprehensive information hub that turns you from a passive consumer of manipulated media into an active, informed participant in democracy. Instead of flooding you with noise, it organizes information to match your needs and interests, with a personalized dashboard that integrates verified news, labeled AI content, fact-checked claims, and direct access to public records and educational resources.

No information overload, no opaque algorithms gaming your attention, no misinformation chaos—just the knowledge you need to be an engaged citizen, delivered clearly and on your terms, right in your pocket.

Summary of Best Practices
How MyVote Implements the Dashboard

Privacy-Preserving Personalization

Your interests and preferences are private.

- *MyVote knows what topics you follow—but this data is not sold, shared, or used for advertising.*
- *Your reading history is not tracked for commercial purposes.*
- *Your political preferences are not profiled or monetized.*
- *You can change interests anytime without judgment or manipulation.*

Algorithm Transparency

Unlike the black-box algorithms of social media, MyVote's information delivery is transparent:

- *You know why you're seeing what you're seeing (it matches your stated interests).*
- *No hidden manipulation to maximize engagement.*
- *No A/B testing different emotional triggers.*
- *No filter bubble—you see opposing viewpoints on issues you follow.*
- *You can see and adjust the algorithm parameters yourself.*

Quality Source Curation

MyVote includes news from verified sources across the political spectrum:

- *Local newspapers and regional journalism.*
- *National news organizations with editorial standards.*
- *International reporting for context.*
- *Verified expert analysis and commentary.*
- *Official government information.*
- *Academic research and policy analysis.*

Sources are evaluated on journalistic standards, not political leanings. Conservative and progressive sources are both included if they meet quality standards.

Fact-Checking Infrastructure

MyVote partners with independent fact-checking organizations:

- *PolitiFact, FactCheck.org, Snopes, and others.*
- *International fact-checking networks.*
- *Academic fact-checking initiatives.*
- *Media literacy organizations.*

Fact-checks are clearly attributed. Multiple sources are cited. You can verify the fact-checkers' work yourself.

Accessibility for All

- **Multiple language support** - *content available in the primary languages spoken in your community.*
- **Screen reader compatible** - *fully accessible for visually impaired users.*
- **Simplified reading levels** - *complex policy explained at multiple reading levels.*
- **Audio versions** - *text-to-speech for all written content.*
- **Visual representations** - *charts, graphs, and infographics for visual learners.*
- **Video content** - *captioned and audio-described.*

Digital Literacy Resources

MyVote includes the standard national K-12 educational curriculum content, as well as:

- *How to evaluate news sources.*
- *Recognizing misinformation tactics.*
- *Understanding media bias vs journalistic standards.*
- *Spotting AI-generated content.*
- *Verifying claims yourself.*
- *Understanding statistics and data visualization.*

- *Critical thinking about political messaging.*

These aren't locked away—they're integrated into the platform where you encounter information.

Chapter 6

Verified Polling and Participatory Budgeting

The Data-Driven Democracy

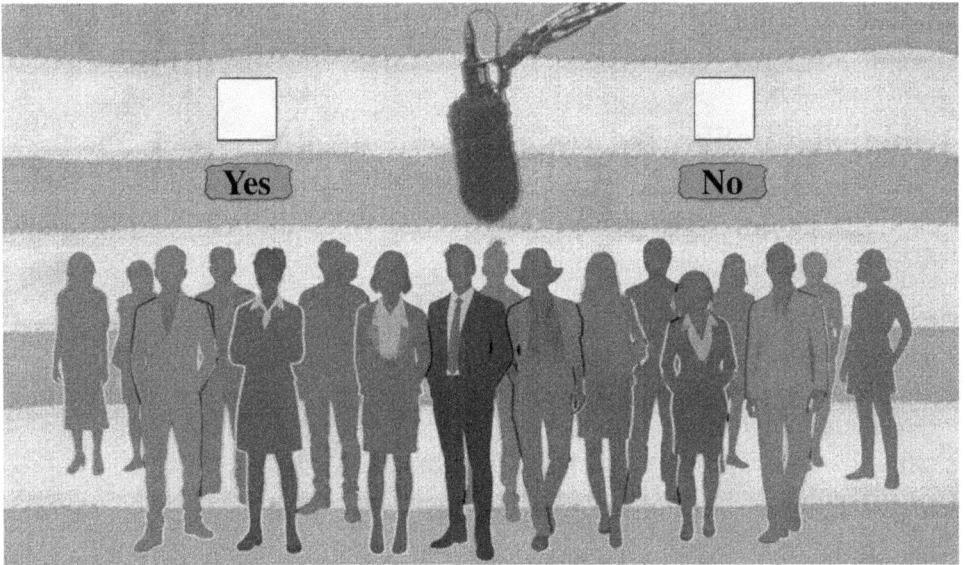

"A modern government of the people, by the people, and for the people needs information from the people to build a better future. So MyVote uses direct polling of every citizen to eliminate the guesswork and deliberate misreading of the public will by one-sided media coverage."

"Instead of paying a lobbyist to buy the attention of your elected representative, your community will now be able to sign official petitions for your grievances, promote real legislation, and vote on policy changes."

"You can even build your own ideal budget to share your priorities with friends and representatives."

- MyVote Demo Video

The Problem: Nobody Can Prove What "The People" Actually Want

Democracy is supposed to be government by the people, but billionaire-owned media manufactures consent by distorting what "the people" actually want. Politicians claim to represent you while citing polls that can't verify respondents. Media claims to report public opinion while cherry-picking data that fits their narrative. Lobbyists claim to speak for citizens while representing corporate interests. Everyone invokes "the will of the people" without any verifiable evidence of what that will actually is.

The information crisis is making democracy impossible.

Traditional Polling Is Fundamentally Broken

Every policy debate begins with competing claims about what Americans want:

"Americans oppose this policy!" says one side, citing a poll.

"Americans support this policy!" says the other side, citing a different poll.

Both polls exist. Both claim scientific methodology. Both purport to measure public opinion. Yet they show completely opposite results.

Why? Because traditional polling of a small group with skewed questions can't verify anything.

The Fatal Flaws

Unknown Respondents: When you read a "national poll of 1,200 likely voters," nobody knows:

- Are they actually registered voters? (Pollsters ask, but can't verify).
- Will they actually vote? (Pollsters create "likely voter models" that are often wrong).
- Do they live where they claim? (No verification possible).
- Are they who they say they are? (For online polls, could be bots, foreign operatives, or the same person answering multiple times).

Pollsters are sampling an unknown population with unknown characteristics. It's like trying to determine the average height of Americans by measuring people walking past a Big-and-Tall store and assuming they represent everyone.

Impossibly Small Samples: 1,200 people out of 160 million voters is 0.00075% of the population. The margin of error is typically ±3-4%, meaning a poll showing 52% support vs. 48% opposition is statistically indistinguishable from a tie—yet media treats it as decisive evidence of public opinion.[40]

Different polls with the same methodology show wildly different results because tiny samples amplify random variation. Yet campaigns, media, and politicians treat these numbers as gospel truth.

Selection Bias: Who actually answers polls?

- **Phone polls:** Mostly elderly people with landlines (young people don't answer unknown numbers)
- **Online polls:** Self-selected respondents who click links (not representative of anyone)
- **Response rates:** Under 6% for phone polls (what do the other 94% think? Nobody knows)[41]

The people who respond to polls are fundamentally different from people who don't—more politically engaged, more opinionated, more available. They don't represent voters. They represent "people who answer polls."[42]

Question Manipulation:

One poll asks "Do you support raising taxes?" 32% say yes, and 68% say no. Another poll asks "Do you support adequately funding schools and infrastructure?" With 71% answering yes, and 29% answering no.

Same policy. Different framing. Opposite results. Pollsters can engineer any outcome they want through question design.[43]

Partisan Polling Shops: Some firms consistently lean toward one party because of methodology choices, question framing, or outright bias. Campaigns commission polls from friendly pollsters, then cite them as "independent polls show…" Media cherry-picks polls that support their narrative. Everyone claims data supports them.[44]

The Result: Nobody knows what citizens actually think because the measurement tools are fundamentally broken.

The "Will of the People" Becomes Whatever Powerful Interests Say It Is

In the absence of reliable polling, "public opinion" becomes a weapon wielded by whoever has the loudest megaphone. Politicians claim mandates that don't exist. Media creates false narratives. Lobbyists manufacture consent with polls that use loaded questions, unrepresentative samples, and biased framing. But they generate headlines like "New poll shows Americans support industry position!" Politicians cite them. Media reports them. Public opinion becomes whatever lobbyists can afford to manufacture, and the media repeats it over and over.

Special Interests Flood Comment Periods:

When federal agencies propose regulations, they're required to accept public comments. Sounds democratic, right?

In reality:

- 90%+ of comments come from organized corporate lobbying campaigns.
- Companies pay firms to submit thousands of identical form letters.
- Real citizens have no idea comment periods exist.
- Agencies cite "overwhelming public opposition" based on manufactured comments.

An example of this occurred when the FCC proposed net neutrality rules. Millions of comments were submitted but analysis found that many came from fake names, stolen identities, and bot-generated text. Nobody knows what actual citizens thought because the process was flooded with fake input.[45]

The New York Attorney General's investigation found that nearly 18 million of the more than 22 million comments the FCC received in its 2017 proceeding to repeal net neutrality rules were fake.[46] The broadband industry funded a secret campaign that generated more than 8.5 million fraudulent comments using stolen identities.[47] An additional 9.3 million fake comments supporting net neutrality used fictitious identities, most submitted by a single 19-year-old college student using automated software.[48]

The Consequence: Democracy becomes government by whoever can most effectively manipulate the perception of public opinion. Not government by actual public opinion—government by the illusion of public opinion, crafted by those with resources to manufacture it.

Representatives Govern Blind, Then Claim Mandate

Representatives are expected to speak for their constituents, but they rarely have a clear, reliable picture of what those constituents actually want. Instead, they rely on a fragmented mix of town halls dominated by the most motivated and extreme voices, phone calls and emails driven by organized advocacy campaigns, skewed polling, conflict driven media coverage, and lobbyist meetings purchased through donations. From this incomplete and easily manipulated picture, representatives make decisions and then claim a popular mandate—insisting they are doing "what my constituents want" without any verifiable, district-wide measure of public opinion.

The same dynamic plays out in national policy debates on healthcare, climate, immigration, and more. Different factions cite different polls that can all be technically "real," yet point in opposite directions, allowing each side to claim the mantle of public opinion without any authoritative, transparent measurement. Media outlets amplify whichever numbers fit their narrative or business interests, and politicians cherry-pick the data that supports their preexisting positions.

The human cost is that policies are passed or blocked based on distorted perceptions of what people want, while representatives can vote against the true majority view in their own districts and still insist they "represented" their constituents—backed by data the public has no meaningful way to verify or challenge.

The Solution: Verified Polling That Can't Be Manipulated, Ignored, or Spun

MyVote implements authenticated, continuous polling of verified voters on every major issue, creating an authoritative, transparent, undeniable measurement of actual public opinion that politicians, media, and lobbyists cannot manipulate or ignore.

Here's how verified polling transforms democracy from guesswork into data…

Authentication: Finally Knowing Who We're Asking

With MyVote's biometric verification we know that every poll respondent is an authenticated voter. This solves traditional polling's fundamental problem: we finally know who we're surveying.

When MyVote says "67% of Congressional District 12 supports policy X," that's 67% of verified registered voters in that district, not 67% of whoever happened to answer a phone call or click a link.

Massive Sample Sizes: Measuring, Not Estimating

Traditional polls survey 800-1,200 people and extrapolate to millions with margins of error of ±3-4%.[49]

MyVote polls can include:

- **Local issues:** Thousands of verified voters (often 5-20% of total electorate)
- **State issues:** Tens of thousands of verified voters
- **National issues:** Hundreds of thousands or millions of verified voters

Instead of a traditional poll of 1,200 people that claims to represent 100,000 people with a margin of error ±3.5%, MyVote verified polls would report the actual views of those 100,000 verified voters with a negligible margin of error. At that level of precision, we are not estimating public opinion. We are measuring it.

Transparent Methodology: Everyone Sees the Same Data

Every MyVote poll shows...

Complete Methodology:

- Sample size (number of verified voters who responded)
- Sampling method (random sample, full population invitation, etc.)
- Response rate (what percentage of invited voters participated)
- Demographic weighting (if applied, and why)
- Confidence interval (statistical certainty level)
- Date conducted
- Exact question wording

Raw Data Available:

- Aggregate results downloadable by anyone
- Demographic breakdowns visible
- Geographic breakdowns by precinct/district/region
- Trend data showing how opinion changes over time

No Proprietary Black Boxes: Unlike traditional polls where methodology is often hidden or vague, MyVote polling is completely transparent. Any statistician, political scientist, or journalist can verify the data independently.

Impossible to Cherry-Pick or Spin

Right now, politicians and media cite whichever poll supports their narrative and ignore the rest:

- "Poll A shows Americans support my position!" (Ignore polls B, C, and D showing the opposite.)
- "Recent polling indicates momentum for our side!" (Cherry-pick the one outlier poll.)

With MyVote as the authoritative source...

There's one definitive measurement of public opinion on any given issue. Not five contradictory polls. One verified, transparent, undeniable data source.

A politician can't cite a friendly pollster when MyVote verified polling shows the opposite. Media can't create false narratives when everyone can see the same verified data. Lobbyists can't commission misleading polls when MyVote's massive, verified sample makes their tiny, biased polls obviously irrelevant.

Continuous Tracking: Not Snapshots, But Movies

Traditional polls are snapshots—they tell you what a tiny group thought on a specific day. MyVote polls are continuous tracking.

How Issues Evolve:

- **January 2025:** MyVote poll shows 45% support for climate legislation, 38% oppose, 17% undecided.
- **April 2025:** After major climate events and policy debate, 52% support, 35% oppose, 13% undecided.
- **July 2025:** After industry campaign against legislation, 48% support, 39% oppose, 13% undecided.
- **October 2025:** After revised legislation addressing concerns, 58% support, 34% oppose, 8% undecided.

This shows:

- How public opinion responds to events.
- How persuasion campaigns impact views.
- How policy changes affect support.
- How opinion crystallizes as people learn more.

Politicians can see which direction opinion is moving and why. Media can report on actual trends, not speculate. Citizens can see how their community's views evolve with information.

Representatives Get Real-Time Constituent Data

Your representative no longer has to guess what you want based on town hall attendance, phone calls from activists, or lobbyist claims. They have verifiable data showing what their actual constituents think.

Example: Congressional District 12 Considers Climate Legislation

MyVote Verified Poll of District 12 Voters:

- **Sample:** 52,847 verified registered voters in District 12 (10.2% of 520,000 total)
- **Margin of error:** ±0.4%

Results:

- Support climate legislation: 61.3%
- Oppose climate legislation: 31.2%
- Undecided: 7.5%

Demographic Breakdown:

- Age 18-35: Support 72%, Oppose 21%
- Age 36-55: Support 63%, Oppose 29%
- Age 56+: Support 51%, Oppose 39%
- Urban voters: Support 71%, Oppose 22%
- Suburban voters: Support 62%, Oppose 30%
- Rural voters: Support 43%, Oppose 48%

This Is Undeniable Information

Rep. Johnson can't claim "my district opposes this legislation" when verified polling shows 61.3% support. She can still vote no—representatives aren't required to follow majority opinion—but she must be honest about what her constituents actually want.

She might explain: "While 61% of my constituents support this in polling, I believe the economic impacts warrant a no vote because…" That's legitimate representative judgment. But she can't lie about what constituents want.

Lobbyists Lose Their Power to Claim Representation

An energy industry lobbyist can't walk into Rep. Johnson's office and claim "my polling shows your constituents oppose this legislation" when MyVote verified polling shows 61.3% support. The lobbyist's poll is obviously garbage when compared to verified data from 52,847 actual constituents.

Lobbying doesn't disappear—lobbyists can still make arguments about policy impacts, economic effects, and implementation challenges. But they can't manufacture fake constituent opposition when real constituent support is verifiable.

Media Must Report Facts, Not Narratives

Cable news loves conflict, so they amplify the most extreme voices and create false impressions of public opinion:

- "Americans are divided on climate policy!" (As they show a shouting match between activists.)
- "Debate rages over proposed legislation!" (As they show paid protesters on both sides.)

With MyVote verified polling...

"MyVote polling shows 61% of District 12 voters support climate legislation, 31% oppose. Rep. Johnson voted no despite clear majority support in her district."

Media can still analyze and debate — but they must start with the facts. They can't create false narratives about divided opinion when verified polling shows clear majority support or opposition.

Journalists can do real accountability reporting:

- "Rep. Johnson voted no on climate bill despite 61% constituent support. Her explanation: [quote]. Do voters find this persuasive?"
- "Districts that voted yes despite constituent opposition: [list]. Districts that voted no despite constituent support: [list]."
- "How representatives vote compared to constituent polling: Analysis shows [data]."

This is real journalism: holding representatives accountable using verifiable data, not speculation about what constituents might want.

Policy Debates Start with Facts, Not Competing Polls

When Congress debates legislation, the conversation changes fundamentally.

Traditional Debate:

- Senator A: "Americans want this policy!" (cites poll from a friendly pollster).
- Senator B: "Americans oppose this policy!" (cites a different poll).
- Media: "Americans are divided on controversial legislation!"

Debate with MyVote Verified Polling:

- Senator A: "MyVote verified polling shows 58% national support, 34% opposition, 8% undecided."
- Senator B: "Yes, but in my state, verified polling shows 42% support, 51% opposition."
- Media: "National polling shows support, but several states show opposition. Here's the breakdown by state with verified data."

The debate shifts from "what do people want?" (unknowable without verified polling) to "given that we know what people want, what's the right policy?" (legitimate policy debate).

Senators can still disagree about whether to follow public opinion—that's the essence of representative vs. direct democracy, but they can't disagree about what public opinion actually is.

Special Interests Can't Flood the Zone with Fake Consensus

Right now, corporations commission polls showing "Americans support" whatever position benefits their bottom line. These polls use loaded questions, tiny samples, and biased methodology—but they generate headlines from friendly news sources.

With MyVote as the authoritative source…

Corporate-commissioned polls become obviously suspect when they contradict verified polling:

Industry Poll: "New poll shows 71% of Americans oppose regulations on industry X!" (commissioned by industry X, with 800 respondents, to the question: "Do you oppose costly regulations that could eliminate jobs?")

MyVote Verified Poll: 847,392 verified voters, 61% support regulating industry X, 32% oppose, 7% undecided. Question: "Do you support federal regulation of industry X to address [specific problem]?"

Media can report: "Industry claims majority opposition, but MyVote verified polling of 847,000 voters shows 61% support. Industry poll surveyed 800 people with a loaded question."

The industry poll becomes obviously irrelevant when compared to verified data from almost a million actual voters.

Participatory Budgeting: You Build the Budget

The federal budget is $6.75 trillion of incomprehensible complexity. MyVote makes it understandable and lets you build your own version to show representatives your priorities.[50]

The MyVote Budget Builder

When a citizen logs into MyVote and accesses the Budget Builder tool they're presented with the current federal budget broken down into major categories:

Mandatory Spending ($4.1 trillion - Social Security, Medicare, Medicaid, etc.):

You see each program with clear explanations:

- **Social Security:** $1.4T - Retirement and disability benefits for 67 million Americans.
- **Medicare:** $1.0T - Healthcare for 65 million Americans over 65.
- **Medicaid:** $600B - Healthcare for 85 million low-income Americans.
- **Other mandatory:** $1.1T - Veterans benefits, federal pensions, nutrition assistance, etc.

Discretionary Spending ($1.7 trillion - Everything else Congress votes on):

- **Defense:** $850B - Military personnel, weapons systems, operations.
- **Education:** $80B - K-12 support, college aid, research.
- **Transportation:** $90B - Highways, transit, rail, aviation.
- **Healthcare Research:** $45B - NIH, CDC, FDA.
- **Energy:** $50B - Research, grid infrastructure, clean energy.
- **Justice:** $35B - FBI, federal prisons, courts.
- **International Affairs:** $50B - Diplomacy, foreign aid.
- **Science & Space:** $40B - NASA, NSF, research.
- **Housing:** $55B - Public housing, rental assistance.
- **Natural Resources:** $45B - National parks, EPA, conservation.
- **Agriculture:** $25B - Farm support, food safety, rural development.
- **All Other:** $335B - Everything else.

You Adjust the Budget

Using simple sliders, you increase or decrease each category. As you move sliders, you see real-time impacts:

- **Increase Education by $20B** → "Would reduce college debt burden, hire 200,000 additional teachers, expand early childhood education."
- **Decrease Defense by $100B** → "Would reduce active duty personnel by 50,000, delay new aircraft carrier, cancel specific weapons programs."
- **Increase Healthcare Research by $10B** → "Would fund 3,500 additional medical research grants, expand pandemic preparedness."

Balance the Budget (or consciously choose deficit spending)

The tool tracks your total spending. If you increase one area, you must either:

1. Decrease another area.
2. Increase revenue (tax options clearly explained).
3. Accept a higher deficit (impact explained clearly).

Revenue Options (if you want to increase spending):

- **Raise corporate tax rate:** Each 1% increase = $9B revenue.
- **Raise top individual rate:** Each 1% increase = $7B revenue.
- **Close specific tax loopholes:** Each explained with revenue impact.
- **New taxes (wealth tax, carbon tax, etc.):** Each fully explained.

See What Others Choose

After building your budget, you see aggregate results:

- "73% of Americans in your district support increasing education funding."
- "58% support reducing defense spending."
- "81% support closing corporate tax loopholes."
- "44% support increasing the top tax rate."

Share Your Budget

- **Post it to your representative:** "This is my budget. Here's why these are my priorities."
- **Share with friends:** "Here's what I'd do with $6.75 trillion. Build yours and let's compare."
- **See your representative's voting record compared to constituent priorities:** "Your rep voted to cut education funding even though 73% of constituents prioritize increasing it."

Representatives Get Aggregate Data

Congressional offices see clear constituent priorities:

- **Average budget from District 12 constituents:** +$40B education, -$80B defense, +$15B infrastructure.
- **Consistency score:** 76% of constituents agree on general priorities even if specific numbers vary.
- **This vs. actual votes:** "You voted for the budget that decreased education funding by $10B when 73% of your constituents wanted increases."

Why Participatory Budgeting Matters

Right now, the budget process happens in darkness. Thousands of pages of incomprehensible line items. Earmarks hidden in unrelated bills. Nobody knows where the money goes, which makes it impossible to hold anyone accountable.[51]

MyVote's Budget Builder changes this.

Transparency: See exactly where every dollar goes, explained in plain language.

Participation: Don't just complain about priorities—show what your priorities are with a complete budget.

Accountability: Representatives can see clear constituent priorities and must explain why they voted differently.

Education: Building a budget forces real tradeoffs. Want more education funding? You have to find the money somewhere—cut defense, raise taxes, or accept higher deficits. This creates informed citizens who understand the complexity of governing.

Common Ground: Real data shows that 73% of your neighbors agree on education funding even if they disagree on other issues. This builds coalitions and reveals consensus that media narratives obscure.

Petition Power: From Symbolic Gesture to Binding Action

MyVote transforms petitions from feel-good clicktivism into official mechanisms with teeth.

Traditional Online Petitions:

- Anyone can sign (no verification).
- No binding effect (government reads them and ignores them).

- Often flooded with bots and fake signatures.
- Result: "We read your petition with 100,000 signatures. Thanks for your input. We're doing nothing."

MyVote Verified Petitions:

- Every signature is a verified, authenticated voter in the relevant jurisdiction.
- Petitions trigger mandatory governmental response when they reach thresholds.
- Geographic verification ensures only affected citizens sign (you can't sign a Springfield petition if you live in Portland).
- Real-time tracking shows exactly how many verified signatures exist.

Petition Thresholds That Trigger Action

Local Petitions (5% of registered voters):

- → Mandatory city council/county board public hearing.
- → Official response required within 30 days.
- → If petition requests specific action, council must vote on it.

State Petitions (3% of registered voters):

- → Referred to appropriate legislative committee.
- → Mandatory public hearings.
- → Committee must vote on advancing to full legislature within 90 days.
- → Detailed report explaining action or inaction.

Federal Petitions (1% of registered voters, ~1.5 million):

- → Referred to relevant congressional committee(s).
- → Mandatory hearings with expert testimony.
- → Committee must draft legislation or issue detailed explanation of why they decline.

Example: Verified Petition Power

Issue: Predatory payday lending in State X charges an average of 400% APR, trapping low-income residents in debt cycles.[52]

Traditional Approach:

- Advocacy groups lobby the legislature.
- The payday lending industry lobbies harder (they have more money).
- The media runs competing stories, inevitably favoring the side with deeper pockets.
- The legislature does nothing (the industry wins).

MyVote Approach:

Day 1: A citizen drafts a "Payday Lending Reform Act" limiting interest rates to 36% APR.

- It is posted to MyVote for community feedback and refinement.
- Other citizens suggest improvements over several weeks.
- Verified subject matter experts (economists, consumer advocates, attorneys) provide input.

Day 60: The refined legislation is ready for a petition drive.

- A citizen launches a MyVote petition with verified signature collection.
- Every signature is from a verified voter in State X.
- Real-time tracking shows the signature count.

Day 120: The petition reaches 157,000 verified signatures (3.1% of the state's registered voters).

- It is automatically referred to the State Senate Banking Committee.
- The committee must schedule public hearings within 30 days.
- The committee must vote on advancing the legislation to the full legislature within 90 days.

Result: The legislature can still vote no—but they must publicly vote on the citizen-drafted legislation that 157,000 verified constituents signed. Industry lobbyists can still argue against it—but they're arguing against documented citizen demand, not abstract policy preferences.

The difference: Citizens don't need lobbyists to get their grievances heard. Verified petitions with documented support force governmental response.

Vote on Policy Changes: Direct Democracy with Safeguards

MyVote enables citizens to vote directly on policy issues (where constitutionally permitted) while maintaining necessary safeguards:

Advisory Referendums (inform representatives but don't directly change law):

- "Should State X expand Medicaid?"
- "Should the U.S. adopt universal background checks for firearms?"
- "Should minimum wage increase to $15/hour?"

Results create political pressure and inform representatives, but don't automatically become law.

Binding Local Referendums (direct democracy on local issues):

- "Should Springfield ban plastic bags?"
- "Should the city issue $50M in bonds for school improvements?"
- "Should downtown be rezoned for mixed-use development?"

If the referendum passes, it becomes law/policy (subject to constitutional constraints).

The Power of Verified Voting

Instead of traditional referendums where turnout is low and results are questionable, MyVote referendums are:

- **Accessible to everyone** (vote from a phone, no need to travel to a polling place)
- **Verified** (every vote from an authenticated voter)
- **Informed** (comprehensive information provided before voting)
- **Transparent** (real-time results, verifiable counts)

Real-World Results: What Verified Polling Achieves

Switzerland's Direct Democracy:

Citizens vote on 10+ national referendums annually.[53] High participation (45-60%) because issues matter and votes count.[54] Policies reflect actual public will, not special interest manipulation.

Result: Extremely stable, prosperous society with high trust in government.[55] Over 60% of Swiss people have high trust in their national government, compared to the OECD average of 39.3%.[56]

Taiwan's vTaiwan Platform:

Digital platform for policy consultation with verified citizen participation. Used for complex issues like ride-sharing regulation and fintech policy.[57] Government must respond to citizen input with documented reasoning.[58]

Result: 80%+ approval for policies developed through the platform.[59] As of 2018, vTaiwan processed 26 digital policy issues and resulted in "decisive government action" in 80 percent of cases.[60]

Iceland's Democratic Experiments:

Digital platforms for citizen input on major policy issues. Verified participation ensures input represents actual citizens.[61]

Result: Higher engagement, better policies, stronger democratic legitimacy.[62] Better Reykjavík platform achieved mass online community participation with 70,000 citizens engaging out of a population of 120,000.[63]

Estonia's Digital Democracy:

E-petitions with verified signatures trigger parliamentary review.[64] Citizens can propose legislation directly. The government must respond with detailed explanations.[65]

Result: Citizens feel heard, trust in government increases.[66] The Rahvaalgatus (citizen initiative) platform has lowered the threshold for policy-making and contributed to a culture of trust between representatives and citizens.[67]

Brazil's Participatory Budgeting:

Porto Alegre pioneered letting citizens directly decide municipal budget priorities. Annual assemblies plus digital participation determine 20%+ of the city budget.

Result: Dramatic improvements in services residents actually wanted, 85% satisfaction rates, reduced corruption.[68]

Why This Matters to You

Voice: Your opinion isn't filtered through media, lobbyists, or activists. You speak directly, and your voice is counted equally with every other verified voter.

Power: Special interests can't manufacture fake consensus when verified polling shows what citizens actually think. Lobbyists lose their ability to claim they represent public opinion when they don't.

Clarity: Politicians can't gaslight you about what "everyone thinks" when verified polling shows exactly what your community thinks. No more "Americans want X" when Americans clearly want Y.

Accountability: Representative can't claim to represent you while voting against clear majority opinion—or if they do, it's documented and undeniable. You can hold them accountable with evidence.

Participation: You don't just choose representatives every two years. You build budgets, sign petitions, vote on policies, and inform every major decision with your verified voice.

Dignity: You're not begging representatives to listen to you. You're not hoping lobbyists will care about your concerns. You're not trusting media to accurately represent your views. You speak for yourself, verifiably and powerfully.

The Bottom Line

Verified polling turns "public opinion" from a political prop into a measurable fact. For too long, numbers about what "Americans think" have been shaped by whoever could afford the polling, the PR, and the lobbyists—letting politicians claim a mandate, media chase clicks, and special interests manufacture consensus without any authoritative, verifiable source of truth. MyVote ends that charade by measuring what real voters actually think, with digital receipts.

With biometric authentication and large sample sizes in the hundreds of thousands or millions, MyVote's verified polls replace guesswork with transparent, auditable data about real constituents. Public opinion stops being an estimate and becomes a shared, undeniable reference point. Representatives can still disagree with their voters, but they can no longer lie about where the district stands; media must report facts instead of narrative-driven polling; lobbyists lose the ability to hide behind cooked-up numbers.

Combined with participatory budgeting, verified petitions, and direct democracy on major issues—with explicit safeguards for minority rights and constitutional limits—this creates the information infrastructure democracy has always needed: accurate, transparent, and impossible to ignore.

Summary of Best Practices
How MyVote Implements Verified Polling

Scientific Methodology

- Random sampling options for quick pulse checks.
- Full population invitations for definitive measurements.
- Stratified sampling ensuring all demographic groups are represented.
- Demographic weighting applied transparently when needed.
- Confidence intervals calculated and published.
- Longitudinal tracking showing opinion changes over time.

Question Design Standards

- Neutral, unbiased wording reviewed by nonpartisan experts.
- Multiple response options beyond just yes/no.
- "Need more information" option always available.
- Exact question wording always published.
- A/B testing to ensure questions aren't accidentally biased.

Information Before Opinion

Before responding to polls, citizens can access:

- Neutral summaries of the issue.
- Arguments for and against from qualified proponents.
- Fact-checks of common claims.
- Expert testimony from verified subject matter experts.
- Fiscal impact analysis.
- Implementation details.

Citizens can skip this if already informed—but it's available and encouraged.

Preventing Gaming

- *One person, one vote (biometric authentication prevents duplicates).*
- *Geographic verification (only affected citizens participate in local polls).*
- *Transparent methodology (anyone can audit the process).*
- *Anti-brigading protections (sudden suspicious spikes in participation trigger investigation).*

Representative Accountability

- *Vote alignment tracking (how often representatives vote with constituent majority).*
- *Response requirements (representatives must respond to verified polling on major issues).*
- *Public dashboards showing how representatives' votes compare to constituent preferences.*

Privacy Protection

- *Individual poll responses are private (nobody sees how you voted).*
- *Only aggregate data is published.*
- *No tracking for commercial purposes.*
- *No selling of data.*
- *Strong encryption protecting everything.*

Chapter 7

Elections and Voting
Democracy's Sacred Moment,
Finally Perfected

"Voting is at the core of democracy, so we're making elections more accessible, informative, and engaging than ever before. Instead of basing your votes on political attack ads and roadside signs for judges and sheriffs, citizens will now be able to view the candidate's biography, platform positions, and political history in a clear and consistent format that includes political donors and other relevant public information, so every voter has all the facts before voting."

"MyVote also ensures a seamless and secure voting experience so your vote is protected and verifiable. Every citizen will now be able to see the results from early voting to final count, with real-time data and analysis, instead of relying on skewed exit polls and paid pundits."

- MyVote Demo Video

The Problem: Democracy's Most Important Moment Is a Disaster

You have the power to choose who governs. It's the most fundamental right in a democracy. And the system makes it nearly impossible to exercise that power responsibly.

The voting information crisis is destroying informed democracy.

Walk into a voting booth tomorrow. You'll face decisions about a President, Senator, Representative, Governor, State Legislators, Mayor, City Council, County Board, School Board, judges at multiple levels, a Sheriff, a County Clerk, a Treasurer, a Coroner, and maybe fifteen ballot initiatives on everything from tax policy to constitutional amendments.

How much do you actually know about these choices?

The Presidential Race: We have all seen thousands of attack ads, cable news shouting matches, and viral social media clips—90% of which are misleading or outright false.[69] We know the scandals. We know the soundbites. Do we know their actual policy positions in detail? Their legislative history? Their donor networks? Probably not.

Down-Ballot Races: We're supposed to vote for State Representative. Do you know who's running? Have you heard them speak? Do you know their positions? For most Americans: no, no, and no.[70] We vote based on party affiliation or name recognition, if we vote at all.

Judicial Races: You're choosing between Judge Sarah Martinez and Judge Robert Chen for Circuit Court. You've never heard of either. There are no debates. No coverage. Maybe you saw a roadside sign. Maybe you recognize one name from somewhere. So you... guess? Skip it? Vote based on which name sounds nicer?[71]

Local Officials: Sheriff, Coroner, County Clerk, Treasurer—these positions have real power over your life. The Sheriff runs the jail and decides law enforcement priorities. The Coroner determines the cause of death in suspicious cases. Do you know anything about the candidates? Have you seen their qualifications? Their track records?

Ballot Initiatives: "Proposition 47: Constitutional Amendment Regarding Municipal Funding Allocation Framework." You have 45 seconds in the voting booth to decide whether to rewrite the state constitution. The official summary is incomprehensible legalese. You saw some TV ads—one side says it'll save schools, the other says it'll bankrupt cities. Both can't be true. Which is it? You have no idea.[72]

Here's what voter information looks like in America today...

Attack Ads Everywhere: Billions spent on negative advertising designed to manipulate, not inform.[73] "Candidate X wants to defund the police!" (No, they proposed shifting 2% of the budget to mental health crisis response.) "Candidate Y will destroy Medicare!" (No, they proposed adjusting cost-sharing for high earners.) Lies work because voters have no easy way to verify claims.

Media Covers Only the Horse Race: News focuses on polling, gaffes, and scandals—not policy substance.[74] "Senator Smith is up 3 points after yesterday's debate gaffe!" What were they debating? What policies did they discuss? Who cares—let's analyze whether the gaffe will move suburban women voters!

Donor Information Hidden: Super PACs and dark money groups spend hundreds of millions influencing elections.[75] We see the ads. We don't see who paid for them. Candidate websites list major donors (if required by law), but the real money flows through untraceable channels. We are voting blind without knowing who owns whom.

Comparison Shopping Impossible: Each candidate has their own website with their own format. One has detailed policy papers. Another has vague slogans. A third hasn't updated their site since the primary. There's no standardized way to compare candidates' positions side by side. We can't even find basic information like "What do these five candidates think about healthcare?" without visiting five different websites with five different formats.

Judges Are Completely Opaque: Judicial candidates can't campaign on how they'll rule because it's unethical, but you're supposed to vote for them anyway.[76] We might find a brief bio if we search hard enough, but we won't find their judicial philosophy, their case history, or their approach to sentencing—nothing that would actually inform our vote.

Ballot Initiative Manipulation: Special interests write confusing ballot language intentionally.[77] They fund misleading campaigns on both sides to create chaos. They time signature gathering to avoid scrutiny. By the time we are in the voting booth, we are so confused we either skip it or vote based on whichever TV ad we saw most recently.

No Verification: When we vote we slide our ballot into a machine but have no idea if our vote was counted correctly. We can't verify it was recorded as we intended. We can't audit the results. We just trust—and in an era of widespread distrust, that's not good enough.[78]

Results Manipulation and Confusion: Exit polls contradict actual results. News networks call races prematurely. Paid pundits spin results to fit their narrative. Foreign actors spread false information about vote counts.[79] By the time actual, verified results are available, conspiracy theories have already taken hold.

The human cost of voter ignorance...

A qualified, experienced judge loses to an unqualified challenger because the challenger had a more recognizable name.[80] A corrupt sheriff gets re-elected because nobody knows about the misconduct lawsuits. A state representative who never shows up to legislative sessions wins again because voters don't know they're absent. A ballot initiative passes that will have disastrous unintended consequences because voters didn't understand what they were voting on.

Meanwhile, special interests celebrate. An uninformed electorate is an easily manipulated electorate. If voters don't know candidates' donor lists, donors can buy influence without accountability. If voters don't understand ballot initiatives, those who write them control the outcomes. If voters can't verify election results, whoever controls the narrative controls legitimacy.

This isn't democracy. This is democratic theater where citizens go through the motions without the information necessary to make informed choices.

The Solution: Complete Voter Information and Transparent Elections

MyVote transforms elections from confusing chaos into clear, informed, verifiable democratic choice by providing comprehensive candidate information, real-time election transparency, and absolute verification of your vote.

Here's how MyVote revolutionizes elections…

Standardized Candidate Profiles: Know Who You're Voting For

Every candidate—from President to County Coroner—has a standardized MyVote profile with complete, verified information in a consistent format.

Basic Information:

- Full legal name and any previous names
- Age and date of birth (verified)
- Current residence (verified to confirm they live in the district)
- Education (degrees verified through institutions)
- Professional background (employment history verified)
- Military service (verified through DOD records)
- Criminal history (if any—pulled from public records)

Political History:

- Previous offices held (with dates and jurisdictions)
- Previous campaigns (wins, losses, vote percentages)
- Party affiliation history (including any party changes)
- Endorsements received (from organizations and officials)
- Campaign finance history from previous races

Policy Positions: For every candidate, standardized questionnaires on key issues.

Example for Congressional Candidate:

- **Healthcare:** "Do you support universal healthcare? If so, what model? If not, what's your alternative?"
- **Climate:** "Do you believe climate change is primarily human-caused? What policies do you support to address it?"
- **Immigration:** "What's your position on border security vs. pathway to citizenship? Specifics?"
- **Taxes:** "Would you raise, lower, or maintain current tax rates? For whom?"
- **Education:** "Federal role in education? Funding priorities? Student debt?"

Every candidate answers the same questions. Every answer is displayed in the same format. You can compare directly, side by side.

Candidate Can't or Won't Answer? That's displayed too:

- "Candidate declined to answer."
- "Candidate did not respond by deadline."
- "Candidate's answer did not address the question."

Silence is information.

Legislative Record (for incumbents):

- Every vote they've cast (with bill summaries)
- Bills they've sponsored or co-sponsored
- Committee assignments and attendance records
- Floor speeches and public statements
- How their votes align with campaign promises
- How their votes align with district constituent polling

Judicial Record (for judges):

- Cases decided (with high-level summaries, anonymized appropriately)
- Judicial philosophy statements
- Bar association ratings[81]
- Any disciplinary actions or complaints
- Reversal rates on appeal (for lower court judges)

- Publications and speaking engagements
- Professional affiliations

Campaign Finance—Complete Transparency:

This is where MyVote changes everything.

Top Donors (Individuals):

- Name, employer, amount donated
- Total donations over time
- Any business relationships with the candidate

Top Donors (Organizations):

- Organization name, industry, amount donated
- Organization's lobbying activities and priorities
- Any contracts or business relationships with the candidate

Super PAC and Dark Money: MyVote tracks spending supporting or opposing the candidate:

- "Americans for Freedom (Super PAC) spent $2.4M supporting this candidate."
- "Better Tomorrow Fund (Dark Money 501c4) spent $1.8M on ads attacking the opponent."
- When possible, MyVote traces dark money to actual sources through investigative journalism partnerships and public filings.[82]

Industry Analysis:

- "This candidate received 47% of campaign funding from the financial services industry."
- "This candidate received 62% of funding from individual donors under $200."
- "This candidate's top donor industry is healthcare, which spent $340K."

Donor Influence Indicators:

- Votes on legislation affecting top donor industries

- Meetings with lobbyists from donor industries (if logged through MyVote)
- Bill sponsorship patterns aligning with donor priorities

We see exactly who's funding whom—and we can judge for ourselves whether that matters.

Public Records Integration:

- Property ownership (public records)
- Business interests (corporate filings, conflict of interest disclosures)
- Legal history (lawsuits, judgments, bankruptcies—all public record)
- Ethics complaints and investigations
- Financial disclosures (required for federal candidates, many state/ local races)

Media and Fact-Checking:

- Verified news coverage of candidate
- Fact-checks of candidate's statements
- Debate performances and transcripts
- Interview archives

Constituent Feedback (for incumbents):

- MyVote constituent satisfaction ratings
- Response time metrics from MyVote messaging
- Casework completion rates
- Office hour attendance and accessibility metrics
- Town hall participation

All in one place. All in the same format. All verifiable.

Ballot Initiative Transparency: Understand What You're Voting On

Ballot initiatives are rewritten from legal jargon into comprehensible information.

Plain-Language Summary:

- "What It Does" (in simple English)
- "Who Benefits" (which groups gain)
- "Who Pays" (cost and funding source)
- "What Changes" (current law vs. proposed law)

Fiscal Analysis:

- Independent analysis of costs (not just from proponents)
- Revenue impact (if it raises or costs money)
- Long-term vs. short-term costs
- Economic impact analysis

Who's Behind It:

- Who wrote the initiative (individuals, organizations)
- Who funded signature gathering (complete donor list)
- Who's funding the campaign (complete transparency)
- What industries or interests benefit

Arguments For and Against:

- Official arguments from both sides
- Fact-checked for accuracy
- Rebuttals to each side's claims
- Analysis of misleading or false claims in campaigns

Expert Analysis:

- Nonpartisan policy experts' assessment
- Legal analysis (constitutional issues, potential challenges)
- Implementation challenges
- Unintended consequences identified

Similar Efforts Elsewhere:

- Has this been tried in other states?
- What were the results?
- Lessons learned from similar policies

Endorsements:

- Which organizations endorse/oppose
- Which politicians endorse/oppose
- Which newspapers recommend yes/no

Example: "Proposition 12: Minimum Wage Increase"

Instead of cryptic legal language, you see...

What It Does: "Raises the state minimum wage from $12/hour to $18/hour over four years ($15 in 2026, $16.50 in 2027, $18 in 2028). After 2028, the minimum wage adjusts annually for inflation."

Who Benefits: "2.4 million workers currently earning under $18/hour would see wage increases. It disproportionately benefits retail, food service, and care workers."

Who Pays: "Small businesses with narrow profit margins may face increased labor costs. Some economists predict modest price increases for consumers (estimated 1.2% for affected goods/services). Some businesses may reduce hours or staffing."

Fiscal Impact: "State costs: +$400M annually in higher wages for state employees. State revenue: +$200M annually from increased income tax. Net cost: $200M annually. Local government costs: +$150M annually."

Who's Behind It: "Written by Service Employees International Union (SEIU). Signature gathering funded by SEIU ($4.2M), other labor unions ($2.1M), and progressive advocacy groups ($800K)."

Campaign Funding:

- Yes Campaign: $18M (Labor unions 72%, individual donations 28%)
- No Campaign: $24M (Business associations 45%, restaurant industry 32%, retail industry 23%)

Expert Analysis: "UC Economics Department analysis predicts 50,000 jobs lost due to increased labor costs, but 2.4M workers see increased earnings totaling $12B annually."

Net effect: reduced income inequality, modest employment reduction.[83]

Similar Efforts: "Seattle raised the minimum wage to $15 in 2014. Results: 3% employment reduction in affected industries, 8% wage increase for low-wage workers, and 1.5% price increases for consumers. Overall, the poverty rate declined 2%."[84]

Now you can make an informed decision because you have actual information, not just competing TV ads.

Secure, Verifiable Voting: Your Vote, Protected and Proven

MyVote implements cryptographically verifiable voting that's simultaneously secure, anonymous, and auditable.[85]

How It Works:

Option 1: Digital Voting (where legally permitted):

1. You authenticate using biometric verification (your face or fingerprint).
2. MyVote confirms your voter registration and eligibility.
3. You receive your ballot—customized for your exact address (correct district races, local measures, etc.).
4. Make your choices with full candidate/initiative information available in-app.
5. Review your complete ballot before submitting.
6. Upon submission, you receive a cryptographic receipt with a unique code.
7. Your vote is encrypted and recorded on a distributed ledger (blockchain-based for immutability).
8. Your receipt code allows you to verify your vote was counted— without revealing how you voted to anyone else.

Option 2: Paper Ballot with Digital Verification:

For in-person or mail voting using traditional paper ballots:

1. Complete your paper ballot (with MyVote candidate information available on your phone for reference).
2. Your ballot is scanned and entered into the system.

3. Receive a receipt with a verification code.
4. Within 24 hours, you can check MyVote to confirm your ballot was received and counted.
5. If there's a discrepancy, you can challenge it with your receipt.

The Security Model:

- **Anonymous but verifiable:** Your vote is separated from your identity cryptographically—no one can see how you voted, but you can verify it was counted correctly[86]
- **Distributed ledger:** No central database to hack—votes are recorded across multiple independent servers.
- **End-to-end encryption:** Your vote is encrypted from the moment you cast it until it's counted.
- **Paper trail:** Even digital votes have a paper backup for audits[87]
- **Open-source code:** Security researchers can audit the entire system.
- **Multiple independent audits:** Random audits by multiple parties to verify counts.

Key Innovation: You can verify your vote was counted without compromising ballot secrecy. Your receipt code shows "Vote recorded in Block #24601" but doesn't show how you voted. Anyone can verify the block exists and was counted. Only you can verify your specific vote within that block.

Real-Time Election Results: Transparency From First Vote to Final Count

No more exit polls. No more paid pundits guessing. No more conspiracy theories about mysterious vote dumps. Just transparent, verifiable, real-time data.

What You See on Election Night...

Live Vote Counts:

- Real-time updates as votes are processed and verified.
- Breakdown by location, time, method (in-person, mail, early voting).
- Visual maps showing results by precinct, county, and district.
- Historical comparison (how is this tracking vs. previous elections?)

Turnout Data:

- Total registered voters.
- Votes cast so far (by method and location).
- Turnout percentage compared to previous elections.
- Demographic breakdowns (age, gender, location—no party data to protect ballot secrecy).

Vote Processing Transparency:

- "47,382 ballots scanned and verified."
- "12,847 ballots in verification queue."
- "234 ballots flagged for signature verification."
- "89 provisional ballots pending eligibility confirmation."

You see the entire process, not just the final number.

Statistical Analysis (Not Pundit Spin)

MyVote provides data-driven analysis:

- "Based on votes counted so far, Candidate A leads by 4.2% with a margin of error ±2.1%."
- "Historically, County X votes typically mirror state results within 1%."
- "Early voting trends suggest higher youth turnout than in 2020 (+8%)."

No opinions. No speculation. Just statistical analysis of actual data.

Anomaly Detection: MyVote's AI flags statistical anomalies for investigation:

- "Precinct 47 shows unusual voting patterns compared to previous elections—flagged for audit."
- "Mail ballot rejection rate in County X significantly higher than the state average—investigation requested."
- "Turnout in District 12 exceeds registered voters—data error, recount initiated."

Transparency prevents fraud and catches errors immediately.

Independent Verification:

- Multiple independent observers can download vote data.
- Academic institutions, news organizations, and watchdog groups can analyze data independently.
- Any discrepancies are flagged immediately and investigated publicly.
- Audit results are published in real-time.

Results Certification: Once all votes are counted and verified:

- Final results published with a full audit trail.
- Every challenged ballot shown with a resolution.
- Statistical analysis confirming results are consistent and valid.
- Certification by election officials (with their verified MyVote accounts).

No more "stop the count" or "keep counting" controversies. The count is transparent from beginning to end.

Real-World Results: What Transparent Elections Achieve

Estonia's Digital Voting Success:

- 46% of voters use digital voting (i-voting) in national elections.[88]
- Zero successful attacks or fraud in over 15 years of operation.
- Voters can verify their votes were counted correctly.
- **Result:** Increased turnout, especially among young voters and those abroad.

Switzerland's Transparency Model:

- Complete campaign finance transparency for all federal elections.[89]
- Standardized candidate information in multiple languages.
- **Result:** High trust in elections (87% confidence in election integrity).

Australia's Mandatory Disclosure System:

- Real-time donation reporting above $15,000.[90]
- Complete transparency of campaign spending.

- **Result:** Dramatically reduced appearance of corruption in elections.

California's Voter Information Guide:

- Standardized candidate statements and ballot measure analysis.[91]
- Fiscal impact analysis for all measures.
- **Result:** Voters report feeling more informed (72% vs. 43% in states without guides).

Colorado's Ballot Tracking:

- Voters receive notifications when their mail ballot is sent, received, and counted[92]
- They can verify their ballot status online.
- **Result:** 97% voter confidence in election integrity despite mail-ballot voting.

Why This Matters to You

Information: You enter the voting booth knowing exactly who you're voting for, what they stand for, who funds them, and what their track record shows. No more guessing based on yard signs.

Confidence: You can verify that your vote was counted. You can see real-time results. You can audit the process yourself. Election conspiracy theories die when everyone can verify everything.

Power: Special interests lose their ability to hide in darkness. When you know who's funding candidates and what their donor relationships are, you can make informed choices about who's really representing your interests.

Participation: When voting is easier, more secure, and more informative, more people participate. Higher turnout means more legitimate representation.[93]

Trust: When elections are transparent from candidate information through vote counting to final certification, trust in democracy increases. We stop fighting about process and start fighting about ideas—as it should be.

The Bottom Line

Democracy dies in darkness, and America's elections are operating in near-total darkness—voters making life-altering decisions about leaders and policies based on attack ads, name recognition, and guesswork rather than actual information about candidates' records, funding sources, and positions. MyVote transforms elections from democratic theater into genuine informed choice by giving every voter standardized, verifiable information about every candidate and ballot measure, complete transparency about who's funding campaigns and why, cryptographically secure voting you can personally verify was counted correctly, and real-time transparent results that eliminate conspiracy theories and restore trust.

When we can see exactly who candidates are, what they've done, who owns them, and verify our vote was counted—while watching transparent, auditable results in real-time—special interests lose their power to manipulate in darkness, and democracy finally works the way it's supposed to with informed citizens making genuine choices about their future.

Summary of Best Practices
How MyVote Implements Election Integrity

Candidate Information Verification:

Every piece of candidate information is verified:

- _Educational records confirmed with institutions._
- _Employment history verified through public records and/or employers._
- _Criminal records pulled from official court databases._
- _Financial disclosures verified against official filings._
- _Legislative votes pulled directly from official government records._

Candidates can dispute information, but must provide evidence. Disputes are shown publicly: "Candidate disputes this characterization. Their response: [...]"

Campaign Finance Real-Time Tracking:

Unlike current systems where donation data is months behind:[94]

- _Donations reported to MyVote within 48 hours (automated for digital payments)._
- _Running totals updated daily._
- _Dark money spending tracked through ad buys and vendor payments._
- _Analysis provided showing funding sources and industry patterns._

Ballot Initiative Review Process:

Before initiatives appear on MyVote:

- _Plain-language summaries reviewed by literacy experts and nonpartisan policy analysts._
- _Fiscal analysis by independent economists._
- _Legal review by constitutional scholars._
- _Fact-checking of proponent/opponent claims._
- _Public comment period for corrections._

Voting Security Protocols:

- **Multi-factor authentication:** Biometric + registered voter verification.
- **Geolocation verification:** Confirms you're voting from your registered address or authorized location.
- **Time-stamped audit trails:** Every vote recorded with timestamp for verification.
- **Air-gapped critical systems:** Most critical vote counting systems not connected to the internet[95]
- **Encryption standards:** Military-grade encryption for all vote data.
- **Regular security audits:** Continuous penetration testing and security reviews.
- **Bug bounty program:** Rewards for security researchers who find vulnerabilities.

Accessibility Provisions:

- **Multiple language support:** Ballots and candidate info in all languages spoken by 5%+ of the jurisdiction[96]
- **Audio ballots:** Full text-to-speech capability.
- **Visual accommodations:** High contrast modes, large text, screen reader compatible.
- **Physical disability accommodations:** Mobile voting units, curbside voting, mail voting.
- **Cognitive disability support:** Simplified language options, assistance available.
- **Rural access:** Voting centers with internet access in every community.

Recount and Audit Procedures:

- **Automatic audits:** Random statistical audits of every election[97]
- **Triggered recounts:** Automatic recount if the margin is within 0.5%.
- **Requested recounts:** Any candidate can request with reasonable cause.
- **Public observation:** Recounts live-streamed and auditable in real-time.
- **Paper trail verification:** Digital votes compared to paper backups.

Chapter 8

Community Empowerment
Democracy, Finally Delivered

"MyVote is the ultimate system for empowering individuals and communities by digitizing democracy."

- MyVote Demo Video

The Democratic Revolution Is Here

For 250 years, American democracy has been limited by the technology of its time. We built a system for an era of quill pens, horseback messengers, and town squares. It was revolutionary then. It's obsolete now.[98]

MyVote is democracy rebuilt for the 21st century.

Not democracy replaced—democracy empowered. Not democracy automated—democracy enhanced. Not democracy simplified—democracy clarified.

What MyVote Gives You

Access: Representatives are no longer hiding behind gatekeepers and donation requirements.[99] They are accountable and accessible.

Voice: We are not shouting into the void every two years and hoping someone hears. We are participating continuously in decisions that affect our lives.[100]

Information: We are not drowning in misinformation and propaganda.[101] We are receiving verified, contextualized, comprehensive information that empowers informed decisions.

Power: We are not passive subjects of governance. We are an active participant with the tools to shape policy, hold representatives accountable, and build the future we want.[102]

Community: We are not isolated and atomized.[103] We are connected to neighbors, organized around shared interests, and collectively powerful.

Dignity: We are not begging for scraps of attention from elected officials too busy with donors and lobbyists.[104] We are citizens with all the rights and powers that should entail—finally realized through technology that makes those rights real.

What MyVote Gives Your Community

Transparency: Government operates in the open. Every decision, every meeting, every vote, every dollar—visible and auditable.[105]

Accountability: Representatives can't hide from their constituents. Their performance is measured, their promises are tracked, and their actions are documented.[106]

Efficiency: Services that took weeks now take minutes. Casework that languished for months gets resolved in days. Information buried on page 17 of a government website is now on your dashboard.[107]

Inclusion: Rural communities aren't locked out. Non-English speakers aren't shut out. People with disabilities aren't excluded. Working parents aren't silenced by meeting times. Everyone participates on equal terms.[108]

Resilience: When disaster strikes, when crises emerge, when decisions must be made quickly—MyVote enables rapid, informed, democratic response instead of slow bureaucratic paralysis.[109]

Trust: When people can see how government actually works, when they can verify everything themselves, when transparency is built into every system—trust rebuilds. Not blind trust in institutions, but earned trust in a smoothly functioning democratic processes.[110]

What MyVote Gives America

A Democracy That Works: Not perfectly. Never perfectly. But functionally. Where citizens are informed, representatives are accountable, and the government actually responds to the governed.[111]

A Citizenry That Engages: Not because they're forced to. Not because celebrities tell them to. But because engagement is possible, productive, and powerful.[112]

A Government That Serves: Not special interests. Not party bosses. Not the highest bidder.[113] But the people who elected it, with systems that make that service verifiable and enforceable.

A Republic We Can Keep: Benjamin Franklin famously replied, when asked what kind of government the Constitutional Convention had created: "A republic, if you can keep it."[114] We're losing it. Not to foreign enemies.

Not to violent revolution. But to slow erosion—corruption, disengagement, manipulation, and disinformation.[115]

MyVote is how we keep it.

The Choice Is Ours

The technology exists. Estonia, Taiwan, Switzerland—they've proven it works.[116] The need is undeniable. Trust in American democracy is at historic lows, misinformation is rampant, and citizens feel powerless.[117]

The only question is whether we have the courage to build the better system of governance that's sitting in front of us waiting to be built.

MyVote doesn't require constitutional amendments. It doesn't require congressional approval. It doesn't require permission from the powerful interests who benefit from the current broken system.

It just requires adoption. By cities, counties, states—one jurisdiction at a time. By citizens demanding better. By representatives brave enough to embrace transparency. By communities ready to reclaim their power.

This is not a partisan vision. Conservatives benefit when government is efficient and accountable. Progressives benefit when citizens have the power to enact change. Libertarians benefit when government must justify every action transparently. Everyone benefits when democracy actually functions the way it's supposed to function.[118]

This is not a utopian fantasy. MyVote doesn't promise to eliminate all problems. People will still disagree. Representatives will still make mistakes. Citizens will still sometimes be wrong. But we'll disagree with better information. We'll make mistakes that are correctable. We'll be wrong about things we can verify and learn from.[119]

This is not a technological solution to political problems. Technology doesn't create virtue. MyVote doesn't make citizens suddenly wise or representatives suddenly selfless. What it does is remove the barriers between the will of the citizens and the action of our government.[120] It gives us the tools. What we build with them is up to us.

The Future MyVote Enables

Imagine waking up tomorrow in a community where...

You can spend five minutes on MyVote over coffee, catching up on the city council meeting, reading about the state legislation affecting your business, and checking that your Social Security issue is being resolved by your representative's office.

You get a notification that your representative is voting today on a bill you care about. You see their vote doesn't match district polling. You send them a message asking why. You get a response within 24 hours explaining their reasoning. You disagree with their reasoning. You share it with your neighbors, who also disagree. You organize.

A local business owner creates a petition to improve downtown parking. You sign it with one click. It reaches the threshold within a week. The city council holds a hearing. You attend virtually because you can't take time off work. The council votes yes because the community support is undeniable. The problem gets solved.[121]

You're invited to participate in building next year's municipal budget. You spend 20 minutes with the budget tool, seeing what different choices would mean. You submit your priorities. So do 4,000 other residents. The city council sees that 78% of residents prioritize parks funding over street widening. They adjust the budget accordingly.[122]

Election Day approaches. You open MyVote and spend an hour reviewing every candidate, reading their platforms, checking their donor lists, watching their town hall responses. You review ballot initiatives with full fiscal analysis and expert commentary. You cast your ballot digitally and receive a verification code. You check later and confirm your vote was counted. The results come in transparently. No controversy. No conspiracy theories. Just verifiable democracy.[123]

This isn't fantasy. This is what's possible when we build democracy's infrastructure for the 21st century.

The Call to Action

If you're a citizen: Demand MyVote in your community. Tell your mayor, your council members, your state representatives—"We want this. We're ready for this. Build it."

If you're an elected official: Be brave. Embrace transparency. Your constituents will reward you for making yourself accessible and accountable. The honest ones always do.[124]

If you're a government official: Start the integration process. Connect your systems to X-Road. Create MyVote accounts for your offices. Document your processes. Transparency makes your job easier, not harder.[125]

If you're a developer: Contribute to the open-source codebase. Audit the security. Improve the systems. This is infrastructure for democracy—it belongs to everyone.[126]

If you're a journalist: Cover Electronic Governance. Explain it. Help citizens understand what's possible. Democracy dies in darkness, but it also dies in complexity. Make it understandable.[127]

If you're an educator: Teach digital literacy and civic engagement using the example of MyVote. Train the next generation to be active, informed, powerful citizens.[128]

If you're a skeptic: Good. Be skeptical. Audit the code. Test the security. Challenge the claims. Democracy needs skeptics. But give it a fair evaluation—compare it to the broken status quo, not to an impossible ideal.[129]

The Stakes

We're at a crossroads. One path leads to continued democratic erosion. Citizens grow more alienated. Misinformation spreads unchecked. Special interests tighten their grip. Turnout declines. Trust evaporates. Democracy becomes a theatrical performance masking oligarchic reality.[130]

The other path leads to democratic renewal. Citizens reclaim power through technology that makes participation possible, information that makes decisions informed, and transparency that makes accountability real.

MyVote is that second path.

It's not the only thing we need. We still need constitutional reforms, campaign finance reform, gerrymandering solutions, a living wage, affordable health care, and a thousand other improvements.[131] But MyVote is what we can build now, with existing technology, without requiring permission from those who benefit from the broken system.

It's democracy's infrastructure for the digital age.

It's how we protect and defend our republic.

It's how we prove that government of the people, by the people, and for the people doesn't have to perish from the earth—it just needs to be rebuilt with the tools of our time.[132]

MyVote: Democracy, Digitized. Community, Empowered. America, Renewed.

The technology exists. The need is undeniable. The choice is ours.

Let's build the democracy we deserve.

The Bottom Line

MyVote connects you directly to representatives, gives you comprehensive candidate information, ensures secure and verifiable voting, and provides transparent election results in real-time. No more attack ads as your only information source. No more voting blind. No more black-box elections. Just complete information, verified votes, transparent counts, and democracy that finally works for the people. This is how we empower communities. This is how we save democracy.

This is MyVote.

We are heading toward the brass ring… are we going to grab it or just shut our eyes and hope for the best?

Chapter 9

Conclusion
(Or Just the Beginning)
Leading Democratic Institutional Reform

"Demand the app today and let your voice be counted so we can build a better future together."

- MyVote Demo Video

How California Can Build MyVote Now

The path forward is more clear than you might think. We are not proposing something utopian or untested. We are proposing something Estonia accomplished over twenty years ago with a fraction of California's resources.[133] They went from Soviet backwater to digital democracy leader while we've been arguing about whether government websites should have password requirements. If Estonia can do this under constant Russian cyberattack[134] with a population the size of San Diego, California can absolutely do it now.

Here's what makes this moment different from every failed e-government initiative we have seen before: California has already done the hard part. The state's been spending billions on digital modernization, cloud migration, API development, and cybersecurity infrastructure through initiatives like Envision 2026.[135] All those investments created the foundation MyVote needs. We are not starting from scratch. We are not trying to convince skeptical bureaucrats that computers exist. We are at the point where California has built the engine, transmission, and wheels— MyVote is just the body that makes it look like a car instead of scattered parts in a garage.

And the political moment? It couldn't be better. Since Trump's second term began, California has positioned itself as an alternative governance model, actively resisting federal overreach while building relationships with other nations and creating parallel regulatory systems.[136] Governor Newsom most likely has national ambitions, the Democratic supermajority wants to prove progressive governance works, and the tech industry is desperate to show that technology can strengthen democracy instead of destroying it. That alignment doesn't come around often.

And hopefully Apple, Silicon Valley's biggest home grown tech company, will decide to take an active role in helping the people of California and America make this system a reality.

Start Small, Start Smart

The biggest mistake would be announcing a massive statewide program that becomes a political target before anyone sees it work. Instead, we should launch MyVote in a single progressive city within the next eighteen months.

Berkeley or Santa Monica would be perfect—tech-savvy populations, reformist leadership, and a medium size that's manageable but meaningful. The pilot scope stays focused on core functions that matter to residents: direct communication with city council members where every interaction is logged and timestamped, authenticated neighborhood forums for local issues, budget visualization so people can see where their money goes and suggest priorities, streamlined access to city services like permits and parking, and better information about local elections.

This isn't about building everything at once. It's about proving the concept works in real life with real people and real elected officials. Within a year, we can show that at least forty percent of eligible voters registered, that city council members are actually responding faster than before, that city hall foot traffic dropped because people can do things online, and that users across all demographics—not just young tech workers—find the system useful. Most importantly, we need zero major security breaches because one hacking incident could kill the whole project.

The pilot will teach us things we can't anticipate from planning documents. How do elected officials actually use direct citizen access—do they engage meaningfully or find ways to game the system? What authentication works for elderly residents without smartphones? What happens in authenticated forums when people still find ways to be jerks despite their names being attached? Where does the X-Road integration[137] hit friction with legacy systems? These lessons are gold. They can prevent us from making expensive mistakes at scale.

Build the Legal Framework While the Pilot Runs

While Berkeley or Santa Monica residents are testing MyVote, the state legislature should pass the California Digital Democracy Act—creating legal status for authenticated digital identity, mandating X-Road interoperability standards across state agencies, establishing privacy protections that actually have teeth, requiring audit trails and transparency, and funding statewide implementation with about five hundred million over three years.

Even better, we should put a constitutional amendment on the 2026 ballot asking Californians directly: "Shall California establish citizens' right to direct digital access to government services and elected representatives?" This accomplishes two things. First, it builds a public mandate that makes

MyVote politically difficult to dismantle later. Second, it forces a statewide conversation about democratic participation that raises awareness and generates grassroots demand.

The California League of Cities should be a partner from day one, not an obstacle we fight later. We create a "digital democracy certified city" program with competitive grants, so municipalities see adoption as an opportunity rather than a mandate. Cities love competing with each other and sharing best practices. Let them create a peer network for implementation lessons while the state provides support.

Scale Through Proof of Success

Once the pilot demonstrates measurable success—and it will, because Estonia's model works—statewide rollout becomes inevitable rather than aspirational. We start with ten additional cities in the first year, mixing sizes, geographies, and political leanings. Los Angeles, San Francisco, San Diego, San Jose, and Sacramento have the capacity to move fast, but we also include some smaller conservative cities to prove this isn't just a progressive bubble concept. We establish regional support hubs so cities aren't isolated when they hit problems.

Year two brings state agencies online. The DMV, Franchise Tax Board, Employment Development Department, and Covered California integrate their services through X-Road. County governments start adopting the system, beginning with the ten largest. We expand to fifty-plus cities. And here's where it gets really interesting—we launch state legislative interfaces that let citizens track bills, contact their representatives directly through logged channels, and submit testimony without needing to take a day off work to drive to Sacramento.

By year three, we're talking about universal access across California. Every municipality, every county, every school district connected through interoperable systems. Full state agency integration so a resident can genuinely access any government service from one authenticated identity. We start rolling out enhanced features like AI-powered constituent services that draft responses to routine questions, predictive policy modeling that shows residents what proposed changes might mean for them, and participatory budgeting at real scale.[138]

The numbers tell the story. Estonia saves 1,345 years of working time annually through digital government—that's 2% of their GDP.[139] California's GDP is $3.9 trillion.[140] Even a 1% efficiency gain is $39 billion per year. The entire five-year implementation might cost $3 billion, but the return on investment is absurd.

Form the Alliance Bloc

This is where MyVote stops being just a California project and becomes a model for democratic renewal. As soon as California proves the system works, we invite Washington, Oregon, Colorado, and Massachusetts to join a Digital Democracy Compact. These states share our political culture and technical capacity. Together, we create mutual recognition of digital identities so your California MyVote login works in Washington and other states. We build shared X-Road infrastructure that cuts costs through economies of scale. We establish an interstate legislative coordination platform where legislators from member states can align on climate policy, immigration, healthcare, or economic issues that cross borders.

This compact becomes an alternative governance structure that delivers better services than the federal government. When residents in five states representing seventy million Americans have direct access to responsive government while people in Texas and Florida are stuck with 1990s-era systems, the pressure for change becomes irresistible. Other states will demand to join. Eventually, the federal government either adopts the model or explains to its citizens why people in California have rights and access that others don't.

The alliance bloc proves that institutional reform works by demonstrating measurable outcomes. Within two years, pilot cities show 30-50% increases in civic engagement. Documented improvement in government responsiveness drops average reply times from weeks to days. Operating costs fall as digital services replace expensive analog processes. Trust in government rises in participating areas while declining everywhere else.[141]

Within five years, statewide adoption demonstrates scalability. The interstate compact shows this isn't just a California quirk—it's replicable. Economic benefits become measurable as GDP growth in participating regions outpaces non-participating areas, just like Estonia's transformation.[142] Young professionals start moving to alliance bloc states instead of fleeing to

Austin or Miami, because engaged democratic participation turns out to matter more than low taxes when you're building a life.

Within ten years, MyVote becomes the model other nations study. The federal government faces political pressure to adopt or explain why it's providing worse service than state governments. Corporate platform power breaks as authenticated public digital spaces replace ad-driven social media for actual civic discourse. And a new generation of leaders emerges who built trust through logged responsiveness rather than donor relationships.

Address the Objections Head-On

People will say this is surveillance state technology. They are wrong, but we need to address the concern directly. X-Road encrypts all data transfers and logs every access so citizens can see exactly who looked at their information.[143] Apple and Android biometric authentication never leaves your device—the government receives an authentication token, not your fingerprint or face scan.[144] The architecture is open source, meaning security researchers and civil liberties groups can audit everything. Strong legal protections get built into the constitutional amendment. Citizens control their own data—we can see all of it, revoke permissions, and delete information.

Here's the reframe: the current system is actual surveillance. Meta and Google monetize our data without meaningful consent or transparency.[145] MyVote gives us control over our information with legal protections that corporations will never offer voluntarily.

The digital divide concern is legitimate but solvable. The state could provide subsidized smartphones to low-income residents at about $300 per device—that's $600 million to reach two million people, which sounds like a lot until you realize it's cheaper than maintaining parallel analog systems forever. Every library, post office, DMV, and community center gets MyVote kiosks with biometric authentication and staff assistance. We build a telephone interface with voice authentication for people who can't or won't use smartphones. Everything gets translated into California's most used languages[146] through AI with human translators for complex interactions. And we maintain paper systems during a five-to-seven-year transition period. Nobody gets forced. Estonia achieved 88% regular usage[147]—we can hit 85% within five years.

The hardest challenge is elected official resistance. Politicians who campaign on transparency and accountability suddenly get nervous when we tell them every constituent interaction will be logged and timestamped. Some officials genuinely fear direct access because it exposes corruption—when constituent opinions conflict with donor preferences, they can't hide behind staff gatekeepers anymore.

But here's the thing: Estonia's officials resisted too, until they saw the political benefits.[148] Representatives who respond quickly build stronger constituent relationships. Early warning on issues lets them get ahead of problems instead of reacting to crises. Direct policy feedback helps them legislate better. We start with reformist officials who actually want accountability and will proudly champion Electronic Governance by publicizing their response times to name and shame laggards, provide staff budget increases tied to engagement, mandate minimum response standards through law, and offer extensive training and AI assistance for routine queries. Make it help them do their jobs better, not just create more work.

Will the federal government try to block this? Probably. They might claim national security concerns and demand backdoor access—we respond with constitutional challenges and public pressure campaigns asking why Washington wants to spy on your conversations with state representatives. They might threaten federal funding—California is already a net donor to the tune of $83 billion annually,[149] so we budget to replace threatened funds and accelerate autonomy. Corporate lawsuits from Meta or Google? We counterclaim antitrust violations and let public outrage build against companies trying to block government services. If federal law attempts to ban state digital identity systems, we face the constitutional crisis directly through mass resistance and Supreme Court challenges.

The point is to position MyVote as resistance to federal overreach and corporate capture. That makes interference attempts politically costly and probably unsuccessful.

Why This Succeeds Where Others Failed

Most digital government initiatives crash for predictable reasons.[150] MyVote avoids every trap.

Failed projects are usually technology solutions looking for problems— governments build digital services because consultants say they should modernize, not because citizens demanded better access. Nobody uses expensive systems that don't solve felt needs. MyVote directly addresses frustrations that residents actually experience: unresponsive government, anonymous online harassment, feeling powerless in democracy, wasted time on bureaucracy, and misinformation overload.

Failed projects get built by contractors optimizing for government processes rather than citizen experience, resulting in unusable interfaces that technically work but nobody wants to touch. MyVote succeeds by hiring actual UX professionals from the tech industry who understand Apple-quality design standards, conducting user testing at every stage with real citizens, building mobile-first because that's where people actually live, and using progressive disclosure so the interface stays simple by default with complexity optional, while baking in accessibility rather than adding it later.

Failed projects have no political champion, so mid-level bureaucrats can't overcome resistance from officials who fear accountability or agencies protecting turf. MyVote could have governor-level championship from Newsom if he sees this as his legacy project, a legislative mandate with actual teeth rather than optional pilots, citizen demand created through ballot measures that force politicians to respond, and interstate compact momentum that becomes too big to ignore.

Failed projects treat security and usability as opposing values, creating risk-averse systems with sixteen-character passwords changed monthly and two-factor authentication that doesn't work. MyVote makes biometric authentication both secure and easy—no passwords to forget, no tokens to lose—while X-Road's architecture has proven secure against nation-state attackers for twenty years.[151] We defend against actual adversaries like Russia and China rather than hypothetical perfect attacks.

Failed projects also create islands of automation where one department builds a digital service that doesn't connect to anything else, so citizens still visit five different websites, re-enter information repeatedly, and carry paper documents between agencies. MyVote solves this by design through X-Road's interoperability layer that connects all systems, enforces the once-only principle where you provide data once and all agencies access it with permission, implements single sign-on with one authenticated identity across all services, and delivers a unified citizen experience that looks like one system even though the backend involves many databases.

The Funding Is There

Five-year total cost runs $2.5-3.5 billion. That sounds like a lot until you break it down. The pilot needs $50-75 million. Statewide legislative authorization takes $25 million. Statewide rollout is the big number at $1.5-2 billion, including infrastructure scaling, agency integration, municipal connections, public access terminals, smartphone subsidies, and training. The interstate phase could add $500 million, while ongoing operations could run $400-600 million annually.

Where does this come from? State budget allocation provides $1.5 billion over five years through the Technology Modernization Fund expansion and reallocation from legacy system maintenance that currently burns $5 billion per year[152] — shifting just 10% to MyVote more than covers it. Federal grants contribute another $500 million through infrastructure funds, fiscal recovery programs, and innovation grants. Philanthropic funding brings in $300-500 million from democracy-focused foundations and tech industry social responsibility budgets — this is how they prove technology can strengthen rather than destroy democracy. A ballot bond measure could add $200 million paid back through long-term savings. And interstate compact cost-sharing could deliver the final $500 million as partner states contribute to shared infrastructure.

The return on investment closes the deal. Estonia's experience shows 2% of GDP in annual savings.[153] California could save $8 billion or more per year at scale. The entire implementation pays for itself within a few years just from efficiency gains, before counting benefits like increased civic engagement, improved policy outcomes, and economic growth from better governance.

The Political Narrative That Wins

Here's how we sell this to Californians across the political spectrum. The core message is simple:

Taking Our Democracy Back

For too long, our government has been captured by wealthy donors, corporate lobbyists, and special interests. We pay our taxes but don't get a voice. Politicians ignore us because they know we can't hold them accountable. Meanwhile, Big Tech companies exploit our data for profit, spread misinformation, and let bots and trolls destroy online discourse.

We are abused workers, not empowered citizens.

MyVote changes everything. It's your platform for e-governance— authenticated, secure, and transparent. Every interaction with your representatives is logged and public. Every government service in one place. Every voice heard, not just the rich and connected.

Estonia did this over twenty years ago. They went from a poor Soviet backwater to one of the world's most advanced democracies. Their people can do anything online in five minutes. They trust their government because they can see exactly what it's doing.

California leads the world in technology. Why are our citizens stuck with twentieth-century government? We can do better. We must do better. MyVote is how we prove democracy still works. It's how we show America and the world that government of the people, by the people, for the people doesn't have to perish from the earth.

Progressive activists hear this as resistance to corporate power and emphasis on direct democracy with climate mobilization potential. Tech workers hear proven solutions they can respect, built on open-source secure architecture with career opportunities in civic tech. Working families hear time savings, convenience, and better government services. Even elderly and conservative voters respond to transparency, accountability, and the promise that participation remains optional with paper systems maintained during transition.

The opposition will come from predictable quarters. Tech libertarians worry about government control, so we emphasize that corporate surveillance is the real dystopia while MyVote gives us control. Privacy advocates fear biometric databases, so we explain no central database exists because biometrics stay on your device. Republicans claim this rigs elections, so we point out authenticated identity prevents fraud better than their proposals.[154] Government unions worry about job losses, so we promise retraining for digital facilitator roles and more staff, just different roles. Corporate platforms resist competition, so we clarify the public square should be public, not corporate property. And cynics will insist government is too incompetent, so we show them Estonia's success and California's technical capacity.

What Happens Next

Everything discussed here is theory until real citizens use the real system and real elected officials respond or don't. The single most important recommendation is this: get the pilot running in one or more cities within eighteen months. Everything else flows from that demonstration effect.

The pilot reveals unforeseen problems we can't anticipate from planning documents. It generates unexpected innovations that improve the design. Most importantly, it overcomes skepticism through visible proof. When citizens in Berkeley or Santa Monica have better government access and responsiveness than citizens everywhere else, the evidence becomes undeniable. When those citizens form new alliances through the platform that solve real problems, the alliance bloc concept proves out. When other cities demand the same system, scalability is demonstrated.

Estonia's success came from pragmatic implementation rather than perfect planning. They built, learned, iterated, and now export their model globally.[155] California can do exactly the same.

Rome fell because institutional reform became impossible within republican structures. The Gracchi brothers tried to reform the late Republic alone and were murdered for it.[156] Success requires networks, not isolated heroes. America faces the same institutional crisis Rome faced—democracy hollowed out while oligarchic power consolidates. MyVote offers a constructive path from late Republic dysfunction to renewed democratic participation, not by recreating what existed before but by building something better suited to our twenty-first-century reality.

The question isn't whether America can avoid Rome's fate. The question is whether California can demonstrate the alternative. When other states and nations adopt California's model despite initial resistance, that demonstrates reform worked. Estonia's model spread to twenty-five countries[157] because it delivered measurable benefits. MyVote must do the same.

This platform could be how we find out. California has the technical capacity, fiscal resources, political will, and innovation ecosystem to lead this transformation. The infrastructure foundation is already built. The political moment is aligned. The proven model exists in Estonia's twenty five-year track record. The need is urgent—climate crisis, water scarcity, and potential national collapse all require unprecedented civic mobilization and trust in government that traditional institutions cannot generate.

The time to start is now. Build the pilot. Make it work. Show the data. Prove that democratic institutional reform isn't utopian dreaming but practical necessity. Demonstrate that effective alliance blocs can form when given proper digital infrastructure. Show the world that government of the people, by the people, and for the people can not only survive but thrive in the twenty-first century.

California can do this. California must do this. The path is clear.

Get out there and demand a modern billionaire-proof digital democracy from your representatives NOW!

Index

A

Accessibility

Accountability

B

Ballot initiatives

Democracy

- direct vs. representative, Ch. 6
- digital infrastructure for, Ch. 5
- information requirements, Ch. 6

Denmark

- digital government, Ch. 5
- NemID system, Ch. 1

Digital divide, solutions, Ch. 7, Ch. 9

Digital identity, Ch. 3

- authentication methods, Ch. 3
- legal framework, Ch. 9
- multi-device security, Ch. 3
- privacy protections, Ch. 3

Direct messaging, to representatives, Ch. 4

Disaster response, integration, Ch. 1

E

Educational resources

- context for current events, Ch. 5
- library integration, Ch. 5
- public records access, Ch. 5

Elections, Ch. 7

- candidate profiles, Ch. 7
- information transparency, Ch. 7
- real-time results, Ch. 7
- security protocols, Ch. 7

Emergency services

- accessibility, Ch. 1
- geolocation features, Ch. 1

- unified access, Ch. 1

Employment support

- job matching services, Ch. 1
- training programs, Ch. 1
- unemployment assistance, Ch. 1

Estonia

- digital democracy model, Ch. 2, Ch. 5, Ch. 6, Ch. 7, Ch. 9
- e-petitions system, Ch. 6
- security against cyberattacks, Ch. 2
- time savings data, Ch. 2, Ch. 9
- X-Road implementation, Ch. 2

F

Fact-checking

- infrastructure, Ch. 5
- misinformation flagging, Ch. 5
- source verification, Ch. 5

Federal agencies, integration of, Ch. 2

Finland, media literacy, Ch. 5

Foreign interference, prevention of, Ch. 3, Ch. 6

Fraud prevention

- authentication role, Ch. 3
- real-time verification, Ch. 2

G

Governance, electronic

- best practices, Ch. 1-7
- implementation phases, Ch. 9
- proven models, Ch. 2, Ch. 5, Ch. 6

Government services

- emergency services, Ch. 1
- healthcare integration, Ch. 1
- housing assistance, Ch. 1
- tax filing, Ch. 1
- veteran services, Ch. 1

H

Healthcare records, integration, Ch. 1, Ch. 2

Housing assistance

- application simplification, Ch. 1
- program consolidation, Ch. 1
- wait list transparency, Ch. 1

I

Iceland

- Better Reykjavík platform, Ch. 4, Ch. 6
- democratic experiments, Ch. 6

Identity theft, prevention of, Ch. 3

Implementation strategy

- California pilot program, Ch. 9
- interstate compact, Ch. 9
- legal framework, Ch. 9
- phased rollout, Ch. 9
- timeline, Ch. 9

Information overload

- current crisis, Ch. 5
- solutions through curation, Ch. 5

Integration, of government systems, Ch. 2

Interstate cooperation, alliance bloc, Ch. 9

J

L

M

MyVote system

- authentication, Ch. 3
- budget builder, Ch. 6
- candidate profiles, Ch. 7
- casework revolution, Ch. 4
- dashboard features, Ch. 5
- emergency services, Ch. 1
- overview, Ch. 1
- petition system, Ch. 6
- verified polling, Ch. 6
- voting security, Ch. 7

N

O

P

Polling, verified, Ch. 6

- authentication requirements, Ch. 6
- massive sample sizes, Ch. 6
- methodology transparency, Ch. 6
- real-time tracking, Ch. 6

Privacy protection

- biometric data, Ch. 3
- constituent communications, Ch. 4
- data control, Ch. 2
- legal safeguards, Ch. 3

Public records

- access integration, Ch. 5
- archive digitization, Ch. 5
- transparency, Ch. 4

R

S

Social Security, service integration, Ch. 1, Ch. 2

South Korea

- K-Government system, Ch. 1
- real-name system lessons, Ch. 3

Special interests, see Lobbying

State integration, with X-Road, Ch. 2

Switzerland

- direct democracy, Ch. 6
- transparency model, Ch. 7

T

Taiwan

- civic tech movement, Ch. 5
- vTaiwan platform, Ch. 4, Ch. 6

Tax filing

- automated pre-filling, Ch. 1
- free direct filing, Ch. 1
- simplification, Ch. 1

Technology companies

- Apple partnership model, Introduction
- role in implementation, Ch. 9

Town halls, virtual accessibility, Ch. 4

Transparency

- algorithmic, Ch. 5
- ballot initiative, Ch. 7
- campaign finance, Ch. 7
- election results, Ch. 7
- government operations, Ch. 4
- legislative activity, Ch. 4

[1] U.S. Department of Veterans Affairs. (2024). *2024 National Veteran Suicide Prevention Annual Report*. The most recent VA report (2022) indicates an average of 17.6 veterans die by suicide daily (down from the often-cited figure of 20-22 from earlier reports). Mission Roll Call. (2025). "The State of Veteran Suicide (2025 Update)." Retrieved from https://missionrollcall.org/veteran-voices/articles/the-state-of-veteran-suicide/

[2] U.S. Department of Housing and Urban Development. (2024). *The 2024 Annual Homeless Assessment Report (AHAR) to Congress*. The 2024 report found more than 770,000 people experiencing homelessness on a single night, an 18 percent increase from 2023. Federal Reserve Bank of Minneapolis. (2025). "Who is homeless in the United States? A 2025 update."

[3] U.S. Department of Housing and Urban Development. (2024). *The 2024 Annual Homeless Assessment Report (AHAR) to Congress*. Nearly 150,000 children experienced homelessness in 2024, a 33 percent increase from 2023. National Alliance to End Homelessness. (2025). *State of Homelessness: 2025 Edition*.

[4] ProPublica. (2017, 2019, 2024). Multiple investigative reports documenting Intuit's and H&R Block's lobbying efforts against free tax filing. NBC News. (2017). "TurboTax, H&R Block Spend Big Bucks Lobbying for Us to Keep Doing Our Own Taxes." In 2016 alone, Intuit spent $2 million on lobbying and H&R Block spent $3 million. ProPublica. (2024). "Inside TurboTax's 20-Year Fight to Stop Americans From Filing Their Taxes for Free."

[5] Fortune. (2023). "Intuit and H&R Block lobby against IRS free tax filing." Analysis shows Intuit, H&R Block, and related advocacy groups spent $39.3 million since 2006 lobbying on free-file and related matters. ProPublica. (2017). "Filing Taxes Could Be Free and Simple. But H&R Block and Intuit Are Still Lobbying Against It."

[6] National Taxpayers Union Foundation. (2023, 2024). *Tax Complexity Reports*. Americans spend 6.5 billion hours and over $260 billion annually (in lost productivity opportunity costs plus out-of-pocket expenses) on tax compliance. Tax Foundation. (2022-2025). Multiple annual reports on IRS tax compliance costs, showing figures ranging from $313 billion to $546 billion depending on methodology and year.

[7] Nordic Institute for Interoperability Solutions (NIIS), "X-Road®," Gaia-X European Association for Data and Cloud, October 31, 2024, https://gaia-x.eu/wp-content/uploads/2024/11/X-RoadNIIS.pdf.

[8] e-Estonia. (2025). Official government sources indicate 99% of public services available digitally. PwC. (2019). "Estonia - the Digital Republic Secured by Blockchain." Report notes that 99% of public services are available digitally and this has saved over 800 years of working time and 2% of GDP annually. Fintech in Baltic. (2024). "Estonia Digitised 99% of Its Public Services With the Aid of Blockchain."

[9] e-Estonia. (2025). "e-Tax" official information page. States that 99% of tax declarations are filed online and the average filing time is 3 minutes. Invest in Estonia. (2023). "Easiness of filing taxes has made it a national pastime in Estonia." Confirms 99% of 2021 tax returns submitted digitally with 3-5 minute filing times.

[10] PR Newswire. (2025). "Estonia Becomes a Fully Digital Nation." Reports that digital signatures save 2% of Estonia's GDP annually and 98% of the population declares income electronically with average 3-minute tax filing. e-Estonia. (2018). "How do Estonians save annually 820 years of work without much effort?"

[11] Government Technology Agency of Singapore. (2022). Official Singpass factsheet indicates more than 4.5 million users, representing 97% of Singapore Citizens and Permanent Residents aged 15 and above. Singapore's population is approximately 5.9 million. Computer Weekly. (2022). "Inside Singapore's national digital identity journey."

[12] Government Technology Agency of Singapore. (2025). Current statistics show Singpass provides access to over 2,000 services across more than 700 government agencies and businesses. Ministry of Digital Development and Information. (2022). "Singpass Factsheet."

[13] Ministry of Digital Development and Information. (2022). "Singpass Factsheet." More than 350 million personal and corporate transactions are facilitated via Singpass annually. World Bank. (2024). Case study reports that over 350 million transactions use Singpass annually to access more than 2,000 public and private sector services.

14 While the exact number of federal legacy systems varies by definition and counting methodology, the U.S. Government Accountability Office (GAO) has extensively documented the federal government's legacy IT crisis. GAO reports from 2016-2023 identify many hundreds of aging federal IT systems across agencies, amounting to thousands of individual components. GAO. (2023). "Information Technology: Agencies Need to Continue Addressing Critical Legacy Systems." Report GAO-23-106821. GAO. (2019). "Information Technology: Federal Agencies Need to Address Aging Legacy Systems." The federal government spends over $100 billion annually on IT, with approximately 80% going to operations and maintenance of existing legacy systems. GAO. (2016). "Information Technology: Federal Agencies Need to Address Aging Legacy Systems." Report GAO-16-468. This report documented systems with components 50+ years old, including Department of Defense systems using 8-inch floppy disks and Treasury systems using 1950s-era assembly language code.

15 FedTech Magazine. (2015, 2018, 2020). Multiple reports on federal legacy IT systems. The 2015 OPM breach exposed personal information of over 20 million government employees partly because COBOL-based legacy systems were too old to support modern encryption tools. Legacy systems pose significant cybersecurity risks due to outdated software, unsupported hardware, and inability to receive security patches. Accenture Federal Services. (2018). "Decouple to Innovate" survey of 185 federal IT executives found that 37% say outdated technology hinders cybersecurity protection, and 46% reported outages involving security breakdowns in legacy systems.

16 e-Estonia official sources report varying figures for time saved through X-Road, reflecting measurements at different times: e-Estonia. (2018). "How do Estonians save annually 820 years of work without much effort?" Reports 820+ years of working time saved annually, noting this represents only 5% of queries with the remaining 95% occurring through automated machine-to-machine exchanges that are harder to measure. e-Estonia. (2024). "Estonia's digital ecosystem is creating a seamless society." Reports X-Road saves Estonians 844 years of working time annually. e-Estonia. (2020). "Interoperability services." Official page states X-Road saves 844 years of working time every year. Cybernetica. (2018, 2022). Multiple case studies report that in 2018, X-Road helped Estonia save 1,407 years of working time. Non-governmental analysis by Future Shift Labs in a 2025 report found that Estonians saved 2,589 working years according to the most recent calendar year, showing continued growth in efficiency gains.

[17] In April-May 2007, Estonia experienced what is widely considered the first major example of state-sponsored cyberwarfare when Russia-based attackers launched coordinated distributed denial-of-service (DDoS) attacks lasting three weeks following Estonia's decision to relocate a Soviet war memorial. Wikipedia. (2025). "2007 cyberattacks on Estonia." Council on Foreign Relations. "Estonian denial of service incident." NATO StratCom Centre of Excellence. (2019). "2007 cyber attacks on Estonia." The attacks targeted government ministries, parliament, banks, media outlets, and telecommunications companies. Almost 60 key websites were offline simultaneously. While some services were disrupted, Estonia's core digital infrastructure and X-Road system continued functioning, demonstrating the resilience of its distributed architecture. The incident led to the creation of NATO's Cooperative Cyber Defence Centre of Excellence in Tallinn. Estonia has since faced additional cyberattacks, including in August 2022 (coinciding with removal of Soviet monuments during the Ukraine war), which government officials described as "the most extensive cyberattack since 2007," though it went "largely unnoticed" due to improved defenses. Euronews. (2022). "Estonia hit by 'most extensive' cyberattack since 2007."

[18] E-solutions in Estonian community pharmacies: A literature review. (2022). Anita Tuula, Kristiina Sepp, Daisy Volmer. *Digital Health*, PMC9301098. "Estonian e-prescription system was first established in 2010 and today, 99.9% of prescriptions are handled online." e-Estonia. (2025). "FACTSHEET E-health in Estonia." States that "99% of all prescriptions to Estonian patients are issued using a digital prescription." The system connects every hospital and pharmacy in Estonia. When prescribing, doctors use computer software to fill an online prescription form that immediately becomes accessible at any pharmacy upon patient request. Patients only need to present ID at the pharmacy, and any insurance benefits automatically apply. PR Newswire. (2024). "Digital Access, AI, and 99% e-Prescriptions: Estonia's e-Health Revolution." Multiple healthcare publications confirm about 100% of Estonian prescriptions are now digital.

[19] e-Estonia. (2025). "e-Business Register." Official government source states: "Since 2011, most companies have been established online using the e-Business Register. The time associated with the registration process has fallen from 5 days to a couple of hours." The e-Business Register allows registration of new companies online using ID card, Mobile-ID, or e-Residency card. Invest in Estonia. (2025). "Starting a company." Confirms that company registration takes approximately 15 minutes to complete the online form, with approval typically within one working day. Multiple business service providers (1Office, Payoneer, Wise, GEOS International, Enty) confirm that Estonian company formation takes 15 minutes to complete online forms with next-day approval. The record for fastest company registration through e-Residency is 15 minutes and 33 seconds. Eesti Firma. (2025). "Company Formation in Estonia."

[20] USC researchers analyzed over 240 million election-related tweets during the 2020 election cycle and found that bots were responsible for millions of tweets, though the studies present varying estimates. USC Annenberg. (2020). "Election 2020 chatter on Twitter busy with bots and conspiracy theorists." Study found thousands of automated accounts posted about the election, with bots believed responsible for "a few million" tweets that reached hundreds of thousands of users. About 13% of users sharing conspiracy narratives were suspected bots. Nature. (2020). "The next-generation bots interfering with the US election." Interview with USC researcher Emilio Ferrara. USC Today. (2016). "Real or not? USC study finds many political tweets come from fake accounts." Earlier 2016 study found bot accounts produced 3.8 million tweets (19% of all election tweets) and 400,000 of 2.8 million users (15%) were bots. PNAS. (2021). "Bots, disinformation, and the first impeachment of U.S. President Donald Trump." Study of 67.7 million impeachment-related tweets found 24,150 bots (1% of users) generated over 31% of all tweets. Among QAnon supporters, bot prevalence was nearly 10%. The exact figure varies by study methodology and time period, but multiple peer-reviewed studies confirm millions of bot-generated political tweets during 2016-2020 election cycles.

[21] Research on online anonymity and harassment shows authenticated or identity-verified settings reduce aggressive behavior. ResearchGate. (2012, 2024). "Anonymity and roles associated with aggressive posts in an online forum." Studies found that "anonymity of online communication minimizes inhibitions and releases individuals from normative and social constraints that regulate behavior in more accountable environments. Unlike authenticated or identity-verified settings such as workplace platforms or professional networks, anonymous or pseudonymous spaces reduce perceived accountability and social visibility, fostering a greater willingness to engage in deviant or aggressive behaviors, including cyberbullying." National Academies Press. (2024). *Social Media and Adolescent Health*, Chapter 7: "Online Harassment." Reports that industry research indicates only 1% of users are consistently toxic, accounting for 5% of harassment incidents. Context and situation are important determinants, making accountability systems more effective than simple banning. Pew Research Center. (2021). "The State of Online Harassment." Found that 41% of Americans have experienced online harassment, with severe forms (stalking, sexual harassment, sustained harassment) doubled since 2014. The report notes that "anonymity that the internet provides" facilitates harassment, though experiences can also involve acquaintances. While specific "70-90%" reduction figures are difficult to verify, multiple studies confirm that accountability mechanisms and reduced anonymity correlate with decreased harassment and aggressive behavior online.

[22] U.S. Senate Report. (2019, 2020). Multiple reports documented Russian operatives using bots and fake accounts to influence 2016 and 2020 elections. Nature. (2020). "The next-generation bots interfering with the US election." Reports Russian operatives used bots to deceive social media users and sway elections. Twitter identified over 50,000 Russian-linked automated accounts tweeting election-related content. Scientific American. (2018, 2024). "How Twitter Bots Help Fuel Political Feuds." Documents bot campaigns seeking to influence Brexit referendum and elections in France, Germany, Austria, Italy, and Catalonia's independence referendum. USC research shows bots targeted prominent accounts to polarize conversations. Washington Post. (2020). "Are 'bots' manipulating the 2020 conversation?" Notes ongoing suspicion of Russian bots and foreign interference across multiple election cycles.

[23] TechCrunch. (2012). "Surprisingly Good Evidence That Real Name Policies Fail To Improve Comments." Reports Carnegie Mellon researchers Daegon Cho and Alessandro Acquisti found South Korea's real-name policy "reduced swearing and 'anti-normative' behavior at the aggregate level by as much as 30%" in certain contexts, though individual users were not dismayed and the policy increased expletives for some demographics. Yale Journal of International Affairs. (2013, 2020). "Real Names and Responsible Speech: The Cases of South Korea, China, and Facebook." Reviews South Korea's real-name verification system from 2005-2012.

[24] Korea IT Times. (2012). "Lessons Learned from South Korea's Real-Name Policy." Korea Communications Commission (KCC) study showed malicious comments accounted for 13.9% of messages in 2007 but decreased only by 0.9 percentage points to 13.0% in 2008, a year after the regulation went into effect. Global Voices. (2012). "South Korea: Internet 'Real Name' Law Violates the Constitution." Confirms KCC findings that real-name policy was largely ineffective, with users moving to overseas websites. Catalysts for Collaboration. "Case study: South Korea's Internet Identity Verification System." Korean Constitutional Court ruled in August 2012 that evidence was insufficient to show a decrease in hateful comments, defamation, and insults following the real-name policy. The system also became a target for massive hacking incidents, including SK Communications' Cyworld breach affecting 35 million Koreans (over half the population). NBC News. (2012). "Real names no longer required online: South Korean court." Constitutional Court said policy discouraged dissent and undermined free speech. Carnegie Endowment. (2021). "The Korean Way With Data." After implementation, "malicious content was reduced on internet bulletin boards as well as comment and reply sections, but the effect was not large."

25 Pew Research Center. (2021). "The State of Online Harassment." Nationally representative survey of 10,000+ U.S. adults found 41% of Americans have personally experienced online harassment. Reports of severe forms (stalking, sexual harassment, sustained harassment, physical threats) have doubled or nearly doubled since 2014. Women are more than twice as likely as men to say harassment was upsetting. 47% of harassed women cited gender as the reason; 42% cited politics. 70% of women view online harassment as a major problem vs. 54% of men. Pew Research Center. (2017). "Online Harassment 2017." Found 66% of Americans have witnessed online harassment directed at others. 18% have been subjected to severe forms. Social media platforms are "especially fertile ground" for harassment. "Anonymity that the internet provides" facilitates these behaviors. Pew Research Center. (2014). "Online Harassment." Found 73% of adult internet users have witnessed online harassment, and 40% have personally experienced it. Young women ages 18-24 experience severe harassment at disproportionately high levels: 26% have been stalked online, 25% experienced online sexual harassment. NTIA. (2024). "Helping Kids Thrive Online: Health, Safety, & Privacy." Reports nearly 16% of U.S. high school students were cyberbullied in 2021. Cyberbullying noted as one of most prevalent preceding risk factors for youth suicide-related behaviors. Right To Be. (2023). "Understanding Online Harassment." Reviews forms of online harassment including doxing, cyberstalking, sexual harassment, and revenge porn. Research shows 57% of people reporting harassment are female, and 46% of women worldwide have received sexist or misogynist comments online.

26 Congressional offices face substantial constituent communication loads. Members often receive thousands of constituent contacts annually across multiple channels (phone, email, mail, in-person). Congressional Research Service. (2024). "Constituent Services: Overview and Resources." CRS Report IF10503. Notes that casework volumes vary considerably by office, with members often receiving "over a thousand constituent requests per year" for assistance with federal agencies. The Regulatory Review. (2024). "Congressional Constituent Service Inquiries." Scholar Kealy notes that "Members often receive over a thousand constituent requests per year on everything from social security benefits and passport issues to complex regulatory issues."

[27] Fireside. (2023). "How Congressional Staffers Can Manage 81 Million Messages From Constituents." In 2022, congressional offices collectively received almost 81 million messages from constituents and sent more than 3.5 million responses. The month with highest inbound emails was June; lowest was October. Member offices received more than 162,000 casework messages from over 128,000 cases in 2022, with average case open time of 40 days. GovFacts. (2025). "How to Contact Your Member of Congress." Every congressional office is a high-volume communications hub where staff systematically open, read, log, and categorize every piece of constituent correspondence. Staffers maintain tallies compiled into daily or weekly reports for senior staff and the member. A form email can be logged in seconds, while physical letters require more handling.

[28] Lobbying spending has grown substantially over the past two decades. Statista/OpenSecrets. (2024). "Total lobbying spending in the United States from 1998 to 2023." Total lobbying spending reached $4.26 billion in 2023, up from $4.11 billion in 2022. Since 2000, lobbying spending has more than doubled. Washington Post. (2022). "Lobbying broke all-time mark in 2021 amid flurry of government spending." The lobbying industry took in $3.7 billion in revenue in 2021 as organizations pressed Congress and the Biden administration over trillions in pandemic spending and rules affecting healthcare, travel, tourism, and other industries. OpenSecrets. (2025). "Federal lobbying set new record in 2024." Lobbying spending has increased by more than $1 billion over the past decade, totaling almost $37 billion since 2015. In each quarter of both 2023 and 2024, federal lobbying spending surpassed $1 billion. 24/7 Wall St. (2022). "Who Spends the Most Lobbying the US Government." U.S. lobbyists raked in a record $3.7 billion in revenue in 2021 from companies, labor unions, and special-interest groups, according to OpenSecrets. This represented 6% growth compared to 2020.

[29] The Regulatory Review. (2024). "Congressional Constituent Service Inquiries." November 14, 2024. Scholar Rachel Potter Kealy's research for the Administrative Conference of the United States (ACUS) found that "Members often receive over a thousand constituent requests per year on everything from social security benefits and passport issues to complex regulatory issues." Congressional Research Service. (2024). "Casework in a Congressional Office." CRS Report RL33209. Casework refers to response or services Members of Congress provide constituents seeking assistance with federal agencies. Common requests involve Social Security, veterans', or other federal benefits; obtaining missing records or payments; or immigration assistance. Each Member office has considerable discretion in how it defines and approaches casework, subject to House or Senate rules and statute.

30 e-Estonia. (2024). "Citizen Initiative portal empowering Estonians in digital age of democracy." A proposal to Parliament requires at least 1,000 signatures; for local government, 1% of residents with voting rights. Upon reaching the signature threshold, the initiative is forwarded to parliament or local government with one click. Assigned to a relevant commission that must discuss it within three months. Within six months, a position on the initiative must be declared. In 2023, Rahvaalgatus broke records with 313,868 digital signatures—more than parliamentary elections (312,182). Estonian Cooperation Assembly. (2020, 2021). "Participatory Parliamentarism: the case of the Estonian Citizens' Initiative Portal." The Estonian Citizens' Initiative portal (ECIP) is regulated by the "Response to Memoranda and Requests for Explanations and Submission of Collective Addresses Act" adopted in 2014. All initiatives with at least 1,000 signatures are deliberated in parliamentary committees. Citizens-Initiative Forum Europa. "The Five Factors of Success Behind Estonia's Citizen Empowering Platform." During seven years, over 500,000 signatures have been given (over a third of Estonian population; EU equivalent would be 200 million signatures). 119 initiatives sent to Parliament, 60 to local municipalities. Several bills have been adopted as direct result, including 2019 prohibition of fur farms (banned July 2023) and 2020 family law amendment legalizing same-sex marriage (June 2023).

31 OECD Observatory of Public Sector Innovation. (2020). "Better Reykjavík." Close to 700 ideas from citizens have been realized by the city, with visible and usable results in all neighborhoods. Over 450 ideas have been processed through agenda setting. Citizens Foundation. (2022). "Better Reykjavík." Close to 700 ideas from citizens have been realized by the city, making all neighborhoods better for citizens to enjoy. 798 citizen projects have been built or are being built from the $3.5m participatory budgeting project since 2011. Nesta. "Better Reykjavík." Almost 60 per cent of citizens have used the platform, and the city has spent €1.9 million on developing more than 200 projects based on ideas from citizens.

32 Citizens Foundation. (2022). "Better Reykjavík." Over 70,000 people have participated out of a population of 120,000 since the site opened. 30,000 registered users have submitted 10,000 ideas and 21,000 points for and against. OECD Observatory. (2020). Over 70,000 people have participated since opening and 27,000 registered users have submitted over 8,900 ideas and 19,000 points for and against. City-REDI Blog, University of Birmingham. (2020). "How Iceland Is Using Digital to Increase Public Participation in Politics." Over 58% of Reykjavík's population have used the site and 12-15% regularly use it—remarkable figures in the world of engagement and consultation.

[33] Citizens Foundation. (2022). "Better Reykjavík." Annual participatory budgeting online voting has attracted participation of around 12.5% of the city's population. In April 2019, the city completed its 8th annual idea generation with 1,053 ideas, 39,000 visitors (37% of voting population), and 5,800 logging in—a new record. Mass participation shows penetration into urban Reykjavík society is well over 50%. Congress.Crowd.Law. "Better Reykjavík." Participation recovered from 5.7% in 2014 to record-setting 12.5% in 2018. To date, 27,000 registered users have submitted over 8,900 proposals. University of Iceland audit in 2015 found just over 40% of Reykjavík residents were pleased with Better Reykjavík. Participedia. "Better Reykjavík: Iceland's Online Participation Platform." The Better Reykjavík project was meant to encourage participation of all age groups, with representation from 16-20 and 61+ age groups. Following 2010 municipal election, city councillors continued leveraging trust-building power of online citizen consultation.

[34] Zuboff, Shoshana. "Facebook's Outrage Algorithms." *The Age of Surveillance Capitalism: The Fight for a Human Future at the New Frontier of Power*. PublicAffairs, 2019. Internal Facebook documents released by whistleblower Frances Haugen in 2021 revealed that Facebook's algorithm weighted "angry" reactions several times higher in some ranking models than "likes" between 2017-2020, prioritizing content that provoked outrage because it generated more engagement. The Washington Post and Nieman Journalism Lab documented that Facebook knew this amplified divisive and misleading content but continued the practice to maximize user engagement and advertising revenue. See also: Wagner, Kurt, and Deepa Seetharaman. "Facebook Executives Shut Down Efforts to Make the Site Less Divisive." *The Wall Street Journal*, May 26, 2020.

[35] UC Davis. "YouTube Video Recommendations Lead to More Extremist Content for Right-Leaning Users, Researchers Suggest." *UC Davis News*, December 13, 2023. https://www.ucdavis.edu/curiosity/news/youtube-video-recommendations-lead-more-extremist-content-right-leaning-users-researchers. Research from 2023 found that YouTube's algorithm can lead users toward more extreme political content, particularly for right-leaning users. A systematic review published in *Cognitive Research: Principles and Implications* (2020) found that 14 of 23 studies implicated YouTube's recommender system in facilitating problematic content pathways. However, research is mixed: some studies show the algorithm favors mainstream content, while others document radicalization pathways. The debate continues, but multiple peer-reviewed studies confirm the algorithm can amplify extreme content under certain conditions.

36 OECD. "Government at a Glance 2025: Estonia." OECD Publishing, 2024. https://www.oecd.org/en/publications/government-at-a-glance-2025-country-notes_da3361e1-en/estonia_89797d03-en.html. In 2023, 83% of Estonian citizens were satisfied with the administrative services they used, significantly above the OECD average of 66%. Estonia scored 0.74 on the Digital Government Index in 2022, compared to the OECD average of 0.61. Earlier surveys showed 76% of entrepreneurs and 67% of citizens satisfied with e-services (2012), with 78% of Estonians regularly interacting with public authorities online by 2017 (far above the EU average of 49%).

37 Open Society Institute - Sofia. "Media Literacy Index 2023." OSIS, 2023. https://osis.bg/?p=4450&lang=en. Finland has ranked first in media literacy and resilience against misinformation among 41 European countries for six consecutive years (2017-2023) with a score of 74 out of 100 points. The ranking is based on quality of education, free media, and high trust among people. Finland's comprehensive media literacy education, integrated throughout the school system since the mid-2010s, with key revisions in 2014 and 2016, has been specifically designed to counter Russian disinformation. Multiple sources confirm Finland's #1 ranking: U.S. News & World Report (2023), CNN (2019), Canada's National Observer (2023), and thisisFINLAND (2022).

38 Hsiao, Yu-Tang, Shu-Yang Lin, Audrey Tang, Darshana Narayanan, and Claudina Sarahe. "vTaiwan: An Empirical Study of Open Consultation Process in Taiwan." SRI International Technical Report, 2018. Taiwan's g0v (gov-zero) civic tech movement has created open-source tools for government transparency and citizen participation since 2012. The vTaiwan platform enables collaborative policymaking on contentious issues through structured digital deliberation. Taiwan's comprehensive approach to digital democracy and fact-checking infrastructure has made it notably resilient to Chinese disinformation operations, despite being a primary target. The country's success demonstrates how transparent information systems and civic technology can protect democratic discourse.

39 European Commission. "eGovernment Benchmark 2023." Publications Office of the European Union, 2023. Denmark consistently ranks among the top performers in the EU's Digital Economy and Society Index (DESI) for digital public services. The country's single digital gateway provides centralized access to all government services and information. Denmark's digital government satisfaction rates exceed 90% in multiple surveys, with nearly universal digital service adoption. The Nordic country's approach combines mandatory digital-by-default services (with exceptions for those unable to use digital tools) with comprehensive user-centered design.

[40] Pew Research Center. (2016). *5 key things to know about the margin of error in election polls*. https://www.pewresearch.org/short-reads/2016/09/08/understanding-the-margin-of-error-in-election-polls/

[41] Kennedy, C., & Hartig, H. (2019). Response rates in telephone surveys have resumed their decline. *Pew Research Center*. https://www.pewresearch.org/short-reads/2019/02/27/response-rates-in-telephone-surveys-have-resumed-their-decline/

[42] Keeter, S., Hatley, N., Kennedy, C., & Lau, A. (2017). What low response rates mean for telephone surveys. *Pew Research Center*. https://www.pewresearch.org/methods/2017/05/15/what-low-response-rates-mean-for-telephone-surveys/

[43] Groves, R. M., Fowler, F. J., Couper, M. P., Lepkowski, J. M., Singer, E., & Tourangeau, R. (2009). *Survey methodology* (2nd ed.). Wiley.

[44] Traugott, M. W., & Lavrakas, P. J. (2008). *The voter's guide to election polls* (4th ed.). Rowman & Littlefield Publishers.

[45] Timberg, C. (2017, November 24). FCC net neutrality process 'corrupted' by fake comments and vanishing consumer complaints, officials say. *The Washington Post*. https://www.washingtonpost.com/news/the-switch/wp/2017/11/24/fcc-net-neutrality-process-corrupted-by-fake-comments-and-vanishing-consumer-complaints-officials-say/

[46] James, L. (2021). *Attorney General James issues report detailing millions of fake comments, revealing secret campaign to influence FCC's 2017 repeal of net neutrality rules* [Press release]. New York State Office of the Attorney General. https://ag.ny.gov/press-release/2021/attorney-general-james-issues-report-detailing-millions-fake-comments-revealing

[47] James, L. (2021). Fake comments: How U.S. companies & partisans hack democracy to undermine the Federal Communications Commission's rulemaking on net neutrality. New York State Office of the Attorney General. https://ag.ny.gov/sites/default/files/reports/oag-fakecommentsreport.pdf

[48] Administrative Conference of the United States. (2021). *Report: Mass, computer-generated, and fraudulent comments*. https://www.acus.gov/sites/default/files/documents/Final%20Report%20on%20Mass%2C%20Computer-Generated%2C%20and%20Fraudulent%20Comments%20%28Final%2006-01-2021%29_0.pdf

[49] Berinsky, A. (2012). Explained: Margin of error. *MIT News*. https://news.mit.edu/2012/explained-margin-of-error-polls-1031

50 Sintomer, Y., Herzberg, C., & Röcke, A. (2008). Participatory budgeting in Europe: Potentials and challenges. *International Journal of Urban and Regional Research*, *32*(1), 164-178.

51 Schick, A. (2007). The federal budget: Politics, policy, process (4th ed.). Brookings Institution Press.

52 Pew Charitable Trusts. (2012). Payday lending in America: Who borrows, where they borrow, and why. Pew Research Center.

53 Swissinfo.ch. (2025). *How Swiss direct democracy works*. https://www.swissinfo.ch/eng/swiss-democracy/how-swiss-direct-democracy-works/89073820

54 Bochsler, D., & Hug, S. (2025). A political-economic analysis of Swiss referendums 1848 to 2022: Turnout, acceptance rates and the double-majority threshold. *Constitutional Political Economy*. https://link.springer.com/article/10.1007/s10602-025-09468-1

55 Fain, C. A. (2011). Modern direct democracy in Switzerland and the American West. Lexington Books.

56 OECD. (2024). *Survey on drivers of trust in public institutions*. Eurac Research. (2024). *Insights on civic participation from Switzerland*. https://www.eurac.edu/en/blogs/eureka/insights-on-civic-participation-from-switzerland

57 Tang, A. (2019). The simple but ingenious system Taiwan uses to crowdsource its laws. *MIT Technology Review*. https://www.technologyreview.com/2018/08/21/240284/the-simple-but-ingenious-system-taiwan-uses-to-crowdsource-its-laws/

58 Hsiao, Y.-T., Lin, S.-Y., Tang, A., Narayanan, D., & Sarahe, C. (2018). *vTaiwan: An empirical study of open consultation process in Taiwan*. https://osf.io/preprints/socarxiv/xyhft/

59 Democracy Technologies. (2024). *Lessons from consensus building in Taiwan*. https://democracy-technologies.org/participation/consensus-building-in-taiwan/

60 European Democracy Hub. (2025). *Digital democracy in Taiwan - Exploring democratic innovations*. https://europeandemocracyhub.epd.eu/exploring-worldwide-democratic-innovations-taiwan/

61 Lackaff, D. (2016). In Iceland, a more direct democracy is already the "new normal." *Medium*. https://medium.com/@lackaff/in-iceland-a-more-direct-democracy-is-already-the-new-normal-91cbe2add46a

[62] Hudson, A. E. (2018). When does public participation make a difference? Evidence from Iceland's crowdsourced constitution. *Policy & Internet*, *10*(2), 185-217. https://doi.org/10.1002/poi3.167

[63] Citizens Foundation. (n.d.). *Your Priorities platform*. https://citizens.is/

[64] e-Estonia. (2024). *Citizen initiative portal empowering Estonians in digital age of democracy*. https://e-estonia.com/citizen-initiative-portal-empowering-estonians-in-digital-age-of-democracy/

[65] Pallo, K.-H. (n.d.). The five factors of success behind Estonia's citizen empowering platform. *Citizens' Initiative Forum*. https://citizens-initiative-forum.europa.eu/citizens-experiences/blogs/five-factors-success-behind-estonias-citizen-empowering-platform_en

[66] Berson, G. (2020). e-Estonia: The ultimate digital democracy? *Medium*. https://medium.com/@geoffrooy/e-estonia-the-ultimate-digital-democracy-f67bc21a6114

[67] Open Government Partnership. (2020). *Rahvaalgatus.ee – yet another e-platform for civic engagement? No, a process of democratic renewal instead!* https://www.opengovpartnership.org/stories/rahvaalgatus-ee-yet-another-e-platform-for-civic-engagement-no-a-process-of-democratic-renewal-instead/

[68] Baiocchi, G. (2003). Participation, activism, and politics: The Porto Alegre experiment. In A. Fung & E. O. Wright (Eds.), *Deepening democracy: Institutional innovations in empowered participatory governance* (pp. 45-76). Verso.

[69] Thorson, E. (2016). Belief echoes: The persistent effects of corrected misinformation. *Political Communication*, *33*(3), 460-480. https://doi.org/10.1080/10584609.2015.1102187; Garrett, R. K., Nisbet, E. C., & Lynch, E. K. (2013). Undermining the corrective effects of media-based political fact checking? The role of contextual cues and naive theory. *Journal of Communication*, *63*(4), 617-637.

[70] Delli Carpini, M. X., & Keeter, S. (1996). What Americans know about politics and why it matters. Yale University Press; Downs, A. (1957). An economic theory of democracy. Harper & Row.

[71] Bonneau, C. W., & Hall, M. G. (2009). In defense of judicial elections. Routledge; Dubois, P. L. (1980). From ballot to bench: Judicial elections and the quest for accountability. University of Texas Press.

72 Matsusaka, J. G. (2004). *For the many or the few: The initiative, public policy, and American democracy*. University of Chicago Press; Lupia, A., & Matsusaka, J. G. (2004). Direct democracy: New approaches to old questions. *Annual Review of Political Science*, 7, 463-482.

73 Geer, J. G. (2006). *In defense of negativity: Attack ads in presidential campaigns*. University of Chicago Press; Franz, M. M., & Ridout, T. N. (2007). Does political advertising persuade? *Political Behavior*, 29(4), 465-491.

74 Patterson, T. E. (2016). News coverage of the 2016 general election: How the press failed the voters. *Shorenstein Center on Media, Politics and Public Policy*, Harvard Kennedy School; Iyengar, S., Norpoth, H., & Hahn, K. S. (2004). Consumer demand for election news: The horserace sells. *Journal of Politics*, 66(1), 157-175.

75 Mayer, J. (2016). *Dark money: The hidden history of the billionaires behind the rise of the radical right*. Doubleday; OpenSecrets. (2024). Outside spending. Center for Responsive Politics. https://www.opensecrets.org/outsidespending

76 American Bar Association. (2003). *Justice at stake: Judicial independence and accountability*. ABA Standing Committee on Judicial Independence; Bonneau & Hall (2009).

77 Gerber, E. R. (1999). The populist paradox: Interest group influence and the promise of direct legislation. Princeton University Press; Broder, D. S. (2000). Democracy derailed: Initiative campaigns and the power of money. Harcourt.

78 Alvarez, R. M., & Hall, T. E. (2008). Electronic elections: The perils and promises of digital democracy. Princeton University Press; Saltman, R. G. (2006). The history and politics of voting technology. Palgrave Macmillan.

79 Allcott, H., & Gentzkow, M. (2017). Social media and fake news in the 2016 election. *Journal of Economic Perspectives*, 31(2), 211-236; Persily, N. (2017). Can democracy survive the internet? *Journal of Democracy*, 28(2), 63-76.

80 Bonneau & Hall (2009); Hall, M. G. (2001). State supreme courts in American democracy: Probing the myths of judicial reform. *American Political Science Review*, 95(2), 315-330.

81 American Bar Association. (2020). Guidelines for reviewing qualifications of candidates for state judicial office. ABA Standing Committee on Federal Judiciary.

82 Center for Responsive Politics. (2024). Dark money basics. OpenSecrets. https://www.opensecrets.org/dark-money; Fang, L., & Becker, A. (2012). How dark money helped Republicans hold the House and hurt voters. *The Nation*.

83 Reich, M., Allegretto, S., & Godoey, A. (2017). Seattle's minimum wage experience 2015-16. Center on Wage and Employment Dynamics, UC Berkeley; Congressional Budget Office. (2021). The effects of a minimum-wage increase on employment and family income.

84 Jardim, E., Long, M. C., Plotnick, R., Van Inwegen, E., Vigdor, J., & Wething, H. (2017). Minimum wage increases, wages, and low-wage employment: Evidence from Seattle. *NBER Working Paper No. 23532*; Zipperer, B., & Schmitt, J. (2018). The "high road" Seattle labor market and the effects of the minimum wage increase. Economic Policy Institute.

85 Benaloh, J., Byrne, M., Eakin, B., et al. (2012). STAR-Vote: A secure, transparent, auditable, and reliable voting system. *USENIX Journal of Election Technology and Systems*, 1(1), 18-37; Chaum, D., et al. (2008). Scantegrity II: End-to-end verifiability for optical scan election systems using invisible ink confirmation codes. *Proceedings of the USENIX/ ACCURATE Electronic Voting Technology Workshop*.

86 Adida, B. (2008). Helios: Web-based open-audit voting. *USENIX Security Symposium*, 17, 335-348; Chaum, D. (2004). Secret-ballot receipts: True voter-verifiable elections. *IEEE Security & Privacy*, 2(1), 38-47.

87 Mercuri, R. (2002). A better ballot box? *IEEE Spectrum*, 39(10), 46-50; National Academies of Sciences, Engineering, and Medicine. (2018). *Securing the vote: Protecting American democracy*. The National Academies Press.

88 Vassil, K., et al. (2016). The diffusion of internet voting: Usage patterns of internet voting in Estonia between 2005 and 2015. *Government Information Quarterly*, 33(3), 453-459; Heiberg, S., Laud, P., & Willemson, J. (2011). The application of I-voting for Estonian parliamentary elections of 2011. *Electronic Voting 2011*, 208, 1-16.

89 Swiss Federal Chancellery. (2023). Political rights: Transparency in campaign financing. https://www.bk.admin.ch/; Transparency International. (2022). Switzerland: Campaign finance regulations.

90 Australian Electoral Commission. (2024). Political donations and financial disclosure. https://www.aec.gov.au/; Gauja, A. (2010). *Political parties and elections: Legislating for representative democracy*. Ashgate.

91 California Secretary of State. (2024). Official voter information guide. https://www.sos.ca.gov/; Lupia, A., & Matsusaka, J. G. (2004). Direct democracy: New approaches to old questions. *Annual Review of Political Science*, 7, 463-482.

92 Colorado Secretary of State. (2024). Ballot tracking system (BallotTrax). https://www.sos.state.co.us/; Stewart, C., & Ansolabehere, S. (2015). Waiting to vote. *Election Law Journal*, 14(1), 47-53.

93 Blais, A. (2000). *To vote or not to vote? The merits and limits of rational choice theory*. University of Pittsburgh Press; Burden, B. C., et al. (2014). Election laws, mobilization, and turnout: The unanticipated consequences of election reform. *American Journal of Political Science*, 58(1), 95-109.

94 La Raja, R. J., & Schaffner, B. F. (2015). *Campaign finance and political polarization: When purists prevail*. University of Michigan Press; Federal Election Commission. (2024). Campaign finance disclosure timeline requirements.

95 Balzarotti, D., et al. (2010). Are your votes really counted? Testing the security of real-world electronic voting systems. *ACM SIGSOFT Software Engineering Notes*, 35(4), 1-5; Springall, D., et al. (2014). Security analysis of the Estonian internet voting system. *Proceedings of the 2014 ACM SIGSAC Conference on Computer and Communications Security*, 703-715.

96 Voting Rights Act of 1965, 52 U.S.C. § 10503 (language assistance provisions); U.S. Election Assistance Commission. (2022). Language accessibility in elections: Best practices guide.

97 Lindeman, M., & Stark, P. B. (2012). A gentle introduction to risk-limiting audits. *IEEE Security & Privacy*, 10(5), 42-49; National Conference of State Legislatures. (2024). Post-election audits.

98 Keyssar, A. (2009). *The Right to Vote: The Contested History of Democracy in the United States*. Basic Books. See discussion of technological constraints on early American democracy in Chapter 1.

99 Lessig, L. (2011). *Republic, Lost: How Money Corrupts Congress—and a Plan to Stop It*. Twelve. Documents how donation requirements create access barriers between citizens and representatives.

100 Fung, A., & Wright, E. O. (2003). Deepening Democracy: Institutional Innovations in Empowered Participatory Governance. Verso. Discusses continuous participation versus episodic electoral engagement.

[101] Lewandowsky, S., Ecker, U. K. H., & Cook, J. (2017). Beyond misinformation: Understanding and coping with the "post-truth" era. *Journal of Applied Research in Memory and Cognition*, 6(4), 353-369.

[102] Pateman, C. (1970). *Participation and Democratic Theory*. Cambridge University Press. Classic work on participatory versus passive models of citizenship.

[103] Putnam, R. D. (2000). *Bowling Alone: The Collapse and Revival of American Community*. Simon & Schuster. Documents the atomization of American civic life.

[104] Schlozman, K. L., Verba, S., & Brady, H. E. (2012). *The Unheavenly Chorus: Unequal Political Voice and the Broken Promise of American Democracy*. Princeton University Press. Examines inequality in political access and representation.

[105] Fung, A., Graham, M., & Weil, D. (2007). *Full Disclosure: The Perils and Promise of Transparency*. Cambridge University Press. Comprehensive analysis of transparency in governance.

[106] Bovens, M. (2007). Analysing and assessing accountability: A conceptual framework. *European Law Journal*, 13(4), 447-468.

[107] Dunleavy, P., Margetts, H., Bastow, S., & Tinkler, J. (2006). *Digital Era Governance: IT Corporations, the State, and e-Government*. Oxford University Press. Documents efficiency gains from digital government services.

[108] Noveck, B. S. (2015). Smart Citizens, Smarter State: The Technologies of Expertise and the Future of Governing. Harvard University Press. Discusses inclusive digital participation mechanisms.

[109] Comfort, L. K., & Kapucu, N. (2006). Inter-organizational coordination in extreme events: The World Trade Center attacks, September 11, 2001. *Natural Hazards*, 39(2), 309-327. On crisis response and democratic resilience.

[110] Grimmelikhuijsen, S., & Meijer, A. (2014). Effects of transparency on the perceived trustworthiness of a government organization: Evidence from an online experiment. *Journal of Public Administration Research and Theory*, 24(1), 137-157.

[111] Dahl, R. A. (1989). *Democracy and Its Critics*. Yale University Press. Establishes criteria for functional democracy.

112 Dalton, R. J. (2008). *The Good Citizen: How a Younger Generation Is Reshaping American Politics*. CQ Press. Examines factors enabling or discouraging civic engagement.

113 Gilens, M., & Page, B. I. (2014). Testing theories of American politics: Elites, interest groups, and average citizens. *Perspectives on Politics*, 12(3), 564-581. Demonstrates disproportionate influence of special interests.

114 Farrand, M. (Ed.). (1911). *The Records of the Federal Convention of 1787* (Vol. 3). Yale University Press, p. 85. McHenry's notes record Franklin's "A republic, if you can keep it" response.

115 Levitsky, S., & Ziblatt, D. (2018). *How Democracies Die*. Crown. Analyzes gradual democratic erosion through institutional degradation.

116 Alvarez, R. M., & Hall, T. E. (2008). *Electronic Elections: The Perils and Promises of Digital Democracy*. Princeton University Press. Includes case studies of Estonia and Switzerland's digital democracy initiatives. For Taiwan: Hsiao, Y. T., Lin, S. W., Tang, A., Narayanan, D., & Sarahe, C. (2018). vTaiwan: An empirical study of open consultation process in Taiwan. SocArXiv Papers.

117 Pew Research Center. (2023). *Public Trust in Government: 1958-2023*. Retrieved from pewresearch.org. Documents historic lows in democratic trust.

118 Inglehart, R., & Welzel, C. (2005). *Modernization, Cultural Change, and Democracy: The Human Development Sequence*. Cambridge University Press. On cross-ideological benefits of democratic functionality.

119 Lindblom, C. E. (1959). The science of "muddling through." *Public Administration Review*, 19(2), 79-88. Classic articulation of incremental, correctable decision-making in democracy.

120 Winner, L. (1980). Do artifacts have politics? *Daedalus*, 109(1), 121-136. Foundational work on how technology shapes but doesn't determine political outcomes.

121 Baiocchi, G., & Ganuza, E. (2017). *Popular Democracy: The Paradox of Participation*. Stanford University Press. Discusses digital petitioning and responsive governance.

122 Cabannes, Y. (2004). Participatory budgeting: A significant contribution to participatory democracy. *Environment and Urbanization*, 16(1), 27-46. Documents participatory budgeting processes and outcomes.

[123] Adida, B., De Marneffe, O., Pereira, O., & Quisquater, J. J. (2009). Electing a university president using open-audit voting: Analysis of real-world use of Helios. *EVT/WOTE*, 9, 10-10. On verifiable voting systems.

[124] Grimmelikhuijsen, S. G., & Welch, E. W. (2012). Developing and testing a theoretical framework for computer-mediated transparency of local governments. *Public Administration Review*, 72(4), 562-571. Evidence on benefits of transparency for officials.

[125] Lips, M., Taylor, J. A., & Organ, J. (2009). Managing citizen identity information in e-government service relationships in the UK. *Public Management Review*, 11(6), 833-856. On government systems integration.

[126] Weber, S. (2004). *The Success of Open Source*. Harvard University Press. On open-source approaches to public infrastructure.

[127] Strömbäck, J. (2005). In search of a standard: Four models of democracy and their normative implications for journalism. *Journalism Studies*, 6(3), 331-345. On journalism's role in democratic functionality.

[128] Bennett, W. L., Wells, C., & Rank, A. (2009). Young citizens and civic learning: Two paradigms of citizenship in the digital age. *Citizenship Studies*, 13(2), 105-120.

[129] Fearon, J. D. (1998). Deliberation as discussion. In J. Elster (Ed.), *Deliberative Democracy* (pp. 44-68). Cambridge University Press. On the productive role of skepticism in democracy.

[130] Wolin, S. S. (2008). Democracy Incorporated: Managed Democracy and the Specter of Inverted Totalitarianism. Princeton University Press. Analyzes democracy as performance versus substance.

[131] Drutman, L. (2020). Breaking the Two-Party Doom Loop: The Case for Multiparty Democracy in America. Oxford University Press. On needed structural democratic reforms.

[132] Lincoln, A. (1863). *Gettysburg Address*. References the closing line "government of the people, by the people, for the people, shall not perish from the earth."

[133] Anthes, Gary. "Estonia: A Model for E-Government." *Communications of the ACM* 58, no. 6 (2015): 18-20. https://doi.org/10.1145/2754951

[134] Ottis, Rain. "Analysis of the 2007 Cyber Attacks Against Estonia from the Information Warfare Perspective." In *Proceedings of the 7th European Conference on Information Warfare*, 163-168. Reading, UK: Academic Publishing Limited, 2008.

135 California Department of Technology. "Envision California 2026: Building a Digital-First State." Sacramento: State of California, 2023. https://cdt.ca.gov/envision-2026/

136 Skelton, George. "Column: California vs. Trump Round 2 is heating up, but will state really 'Trump-proof' itself?" *Los Angeles Times*, January 16, 2025.

137 Pappel, Ingrid, Ingmar Pappel, Valentyna Tsap, and Dirk Draheim. "Exploring X-Road as a Platform for Transnational Digital Government Services." In *Electronic Government and the Information Systems Perspective*, edited by Andreja Pucihar et al., 213-225. Cham: Springer, 2020.

138 Sintomer, Yves, Carsten Herzberg, and Anja Röcke. "Participatory Budgeting in Europe: Potentials and Challenges." *International Journal of Urban and Regional Research* 32, no. 1 (2008): 164-178.

139 e-Estonia. "We Have Built a Digital Society and So Can You." Tallinn: e-Estonia Briefing Centre, 2024. https://e-estonia.com/

140 U.S. Bureau of Economic Analysis. "Gross Domestic Product by State, 4th Quarter 2024 and Annual 2024." Washington, DC: U.S. Department of Commerce, 2025.

141 Pew Research Center. "Public Trust in Government: 1958-2024." Washington, DC: Pew Research Center, 2024. https://www.pewresearch.org/politics/2024/06/24/public-trust-in-government-1958-2024/

142 Kattel, Rainer, and Veiko Lember. "Public Sector Innovation and Economic Growth: The Case of Estonia." In *Handbook on Innovation in Public Services*, edited by Victor Bekkers et al., 456-471. Cheltenham, UK: Edward Elgar Publishing, 2018.

143 Nordic Institute for Interoperability Solutions (NIIS). "X-Road: Technical Architecture." Tallinn: NIIS, 2024. https://x-road.global/

144 Apple Inc. "Face ID and Touch ID Security." Apple Platform Security Guide. Cupertino, CA: Apple Inc., 2024.

145 Zuboff, Shoshana. The Age of Surveillance Capitalism: The Fight for a Human Future at the New Frontier of Power. New York: PublicAffairs, 2019.

146 U.S. Census Bureau. "Language Spoken at Home: California." American Community Survey 5-Year Estimates, 2023. Washington, DC: U.S. Census Bureau, 2024.

147 Vassil, Kristjan, Mihkel Solvak, Priit Vinkel, Alexander H. Trechsel, and R. Michael Alvarez. "The Diffusion of Internet Voting. Usage Patterns of Internet Voting in Estonia Between 2005 and 2015." *Government Information Quarterly* 33, no. 3 (2016): 453-459.

148 Margetts, Helen, and André Naumann. "Government as a Platform: What Can Estonia Show the World?" University of Oxford Policy Brief. Oxford: Oxford Internet Institute, 2017.

149 New York State Comptroller. "Balance of Payments: How Much Does Your State Receive Back From Washington?" Albany: Office of the New York State Comptroller, 2024.

150 Goldfinch, Shaun. "Pessimism, Computer Failure, and Information Systems Development in the Public Sector." *Public Administration Review* 67, no. 5 (2007): 917-929.

151 Parsovs, Arnis. "Practical Security Analysis of E-Government Systems." In *Financial Cryptography and Data Security*, edited by Matthew Brennan et al., 315-327. Berlin: Springer, 2017.

152 California Legislative Analyst's Office. "The 2024-25 Budget: Information Technology." Sacramento: LAO, 2024. https://lao.ca.gov/

153 e-Estonia. "We Have Built a Digital Society and So Can You." Tallinn: e-Estonia Briefing Centre, 2024. https://e-estonia.com/

154 National Conference of State Legislatures. "Voter ID Laws." Washington, DC: NCSL, 2024. https://www.ncsl.org/elections-and-campaigns/voter-id

155 Kitsing, Meelis. "Success Without Strategy: E-Government Development in Estonia." *Policy & Internet* 3, no. 1 (2011): 1-21.

156 Appian. *The Civil Wars*. Translated by John Carter. London: Penguin Classics, 1996. Book I, sections 9-17.

157 Nordic Institute for Interoperability Solutions. "X-Road Community: Members and Participants." Tallinn: NIIS, 2024. https://x-road.global/community

www.ingramcontent.com/pod-product-compliance
Lightning Source LLC
Chambersburg PA
CBHW071346290326
41933CB00041B/2761